IT HAPPENED IN MINNESOTA

It Happened In Series

IT HAPPENED IN
MINNESOTA

Darrell Ehrlick

TWODOT®

GUILFORD, CONNECTICUT
HELENA, MONTANA

AN IMPRINT OF THE GLOBE PEQUOT PRESS

A · TWODOT® · BOOK

Copyright © 2008 Morris Book Publishing, LLC

Text design by Nancy Freeborn
Map by M. A. Dubé © 2008 Morris Book Publishing, LLC

Front cover photo: Minnesota State Fair, balloon vendor. Library of Congress, LC-USZ62-46589
Back cover photo: Flour mills and train on Great Northern Viaduct, Minneapolis, ca. 1903. Library of Congress, LC-USZ62-99818.

Library of Congress Cataloging-in-Publication Data is available on file.

ISBN 978-0-7627-4332-2

Printed in the United States of America

10 9 8 7 6 5 4 3 2

For Angela,
my inspiration and my true companion

WINNIPEG

MANITOBA

Red River

Lake of the Woods

ONTARIO

50 100 MILES
50 100 KILOMETERS

Rainy River

Rainy Lake

Seine River

Lac la Croix

Saganaga Lake

Upper Red Lake

Lower Red Lake

Vermilion Lake

RED RIVER VALLEY

Red River

East Grand Forks

ADA

Lake Winnibigoshish

EVELETH

VIRGINIA

Lake Superior

MOORHEAD

DULUTH

NORTH DAKOTA

PELICAN RAPIDS

MILFORD MINE

Mille Lacs Lake

FERGUS FALLS

BRECKENRIDGE

LITTLE FALLS

KENSINGTON

HINCKLEY

St. Croix River

BROWNS VALLEY

MILLE LACS

WISCONSIN

CROSBY

Big Stone Lake

Mississippi River

ST. CLOUD

MINNEAPOLIS

ST. PAUL

Chippewa River

BLOOMINGTON

SOUTH DAKOTA

Minnesota River

LE SUEUR

NORTHFIELD

LAKE CITY

NEW ULM

MANKATO

PLAINVIEW

MINNESOTA CITY

PIPESTONE

ROCHESTER

WINONA

Mississippi River

AUSTIN

IOWA

Wisconsin River

MINNESOTA

CONTENTS

CONTENTS

PREFACE

Growing up in Montana, my aunt taught my sister and me a song that poked fun at Minnesota. I don't think she had anything against Minnesota; it was just the state a couple doors down from Montana, and we snickered a little every time we sang the tune, like making fun of the neighbor kids on the other side of the street. The tune ended with a rousing, "Minn-ee-sota, yah sure," sung with all the Scandanavian accent a bunch of Montana kids could muster.

Fast forward a decade or so. I visited Concordia College in Moorhead, Minnesota, on what was the coldest day on record—so cold, much of the town shut down. I couldn't tell much about the campus; it was covered in snow, and the cold was so bitter it stung just to open your eyes. I looked around and thought to myself skeptically, "Minnesota, yah sure."

Sometime around my sophomore year in college a couple of friends and I went out to "the lake." When you have ten thousand of them, it doesn't really matter which one and I've long since forgotten which one. I never got why Minnesotans are so taken with their lakes until I happened to fall asleep one evening on an Adirondack chair on the lakeshore. The sound of the waves gently slapping the shore, coupled with a late spring breeze made for one of the most peaceful night's sleep I've ever had. When I awoke I remember thinking something to the effect of "Minnesota, yah sure."

Only this time the tone had changed.

I began to understand what the state was really all about. Minnesota meant the bitter cold of a Moorhead winter when all there seemed to be for entertainment was tossing boiling water into the air only to watch it turn to snow before hitting the ground. Minnesota had something to do with the red and green fireflies flying against a pitch black summer night on a Mower County farm. It's about home-run hankies, going up to Dulut', potlucks, hot dishes, and keeping the coffee pot on. Minnesota, yah sure.

When I convinced a Minnesota farm girl to say "yah, sure" at the altar, I inherited a native state by marriage. The more I studied the North Star State, though, the more I realized I inherited a state whose history I love.

For all the teasing Minnesota gets about being homogenous and fairly reserved, the stereotype just doesn't match up with history or reality. In fact, Minnesota's history and geography seem to mirror the diversity of the nation.

If you want wars, disasters, inventions, scandals, tragedies, spectacles, and triumphs, you need look no further than Minnesota. From the rich farmland, to the boundary waters, to the Iron Range, along the Mississippi River valley, and through the rolling hills, it's sometimes hard to believe you're in the same state.

We'll never know exactly what explorers like Father Louis Hennepin or even Zebulon Pike first uttered when they saw Minnesota, even though they left diaries. Even applying the term *Minnesota* to what was a vast expanse of wilderness doesn't quite match up. But we do know that people were stunned by the beauty, diversity, and ruggedness of the northland. We don't know what words crossed their lips, but in my mind, I have to believe it was something like, "Minnesota, yah, sure."

ACKNOWLEDGMENTS

I wouldn't be very Minnesota-nice if I didn't acknowledge and thank all those who helped me put this book together. But, before I begin, I also want to take a moment to say that any errors or oversights in this book are strictly the fault of the author and in no way should reflect upon the people named in this list of acknowledgments, whose only fault is associating with the likes of me.

I want to thank my editor, Pat Straub, and publisher, The Globe Pequot Press, for giving me the opportunity to do two things I love— write and read history.

I also want to thank the folks at Winona State University's Darrell Krueger Library who were generous with their time and tireless in their efforts to track down materials I needed. A thank-you also needs to be extended to the taxpayers of Minnesota, who, despite having to live with so many program cuts recently, have still seen fit to fund a wonderful library interloan program which makes virtually any book in the state no farther than a trip to the local library.

Walt Bennick, archivist for the Winona County Historical Society, showed his historical prowess searching through files and in other obscure places to find wonderful pieces of the past. Walt, you belong beside all those other valuable treasures you have so lovingly helped preserve.

I would like to thank my boss and company, Rusty Cunningham and Lee Enterprises for letting me keep my day job and encouraging me to pursue my passion of writing about history. It is an inspiring experience to work for a company that sees such value in local history, and I cannot imagine anyone being more supportive or a better teacher than Rusty.

Thanks to Jerome Christenson for his careful attention to historical detail and his contagious passion for Minnesota history.

I need to thank my parents, Darrell and Sally Ehrlick, for the work ethic that they gave me. Of all the things I used while writing this book, their gifts of hard work and discipline were used most.

I want to thank my wife for being my biggest fan and supporter; and for not using the good sense God gave her to run the other way when I asked her to dinner for the first time. As I have continued to say, Angie is the most patient, understanding person I know. And, her patience during the many hours of research, writing, and editing was simply amazing.

Finally, I thank my sister, Melannie (also known to others close to her as "Sweetie"). Without her accidentally finding this opportunity for me, I would not have had the chance to write this book. While it is usually younger siblings who look up to the older ones, I admire her. And I should probably—in the permanence of ink and binding—apologize here for all the transgressions and torments of our youth (especially that macaroni-and-cheese incident). That you still speak to me shows amazing compassion and character.

HENNEPIN'S WILD ADVENTURES

- 1680 -

CLOAKED IN A BLACK PRIEST'S CASSOCK AND trawling through the muddy waters of the Mississippi River, Father Louis Hennepin and his two traveling companions went in search of souls to save and the riches of the New World. But, hiding in tall grasses were some not-so-friendly companions. Hennepin was being watched.

In an instant, a traveling war party of Issati Sioux surrounded the boat, ready to kill Hennepin and his companions. Hennepin had prayed to meet American Indians, but at that moment it looked more like he was going to meet his maker.

One-hundred twenty Indians in thirty-three canoes charged the small boat. Arrows flew past a stunned Hennepin and his crew, who were part of René-Robert Cavelier de La Salle's expedition. As tribal elders drew closer to the pale foreigners, they saw Hennepin holding a peace calumet in his hand. The elders quickly stopped the younger warriors from killing the explorers.

The Sioux had been pursuing their enemies, the Miamis, when they encountered Hennepin, who became the bearer of bad tidings. Hennepin explained that the Sioux were too late—the Miamis had already fled the area. Not wanting their pursuit of the Miamis to be a total loss, the Sioux decided to take Hennepin and his small crew as a consolation prize.

While Hennepin had dreamed of coming to the New World when he was in Europe, he hadn't exactly dreamed of being captured by Indians.

The Sioux found Hennepin sufficiently stocked with supplies, which made their fruitless trip after the Miamis worthwhile. The different chiefs discussed what was to be done with Hennepin. The warriors wanted to introduce Hennepin and his companions to a tomahawk. One chief, Aquipaguetin, wanted to kill the trio because he had not gotten to fight with the Miamis to avenge his son's death.

"[Aquipaguetin] wept through almost every night him [whom] he had lost in war, to oblige those who had come out to avenge him, to kill us and seize all we had, so as to be able to pursue his enemies," Hennepin wrote two years later in *A Description of Louisiana,* after having safely returned to Europe. "But those who liked European goods were much disposed to preserve us, so as to attract other Frenchmen there and get iron, which is extremely precious in their eyes; but of which they knew the great utility only when they saw one of our French canoemen kill three or four wild geese or turkeys at a single gun shot, while they can scarcely kill even one with an arrow."

The Sioux decided to spare the lives of Hennepin and his companions and took the captives with them, beginning weeks of travel, passing Lake Pepin, which Hennepin called the "Lake of Tears" because of Aquipaguetin's wailing. The chieftain's mood vacillated between compassion and revenge and Hennepin records in

A Description of Louisiana that Aquipaguetin often tried to convince the warriors to kill the explorers.

To gain favor with the mercurial chief, Hennepin tried to appease him with gifts of tobacco and twenty knives. This attempt prompted Aquipaguetin to ask his warriors one by one what should be done with the captives, either accept their peace offerings or behead them. Just as Hennepin and his men had bowed their heads, Narrhetoba, a younger chief, stepped in and led the hostages to his camp. Narrhetoba took arrows and broke them "showing us by this action that he prevented their killing us," Hennepin recorded.

Just days after being taken captive, the captors and the captives stopped about fifteen miles south of what would later become St. Paul, where they left their canoes and traveled to where the Rum River flows from Mille Lacs.

The journey was long and treacherous as the Sioux and their captives hiked through swamps, forests, and prairies. The journey exhausted Hennepin so much that he "often lay down on the way, resolved to die there."

Hennepin's protest was met with an equally creative solution: The Sioux set fire to the prairies so, as Hennepin puts it, "we had to advance or burn."

When the party arrived in the Sioux village, things did not get better. Unable to speak the language, the three captives arrived to witness a ceremony they thought was certainly the prelude to their own sacrifice. Instead, it was a peace offering, and Hennepin records dining on wild rice seasoned with whortleberries.

Life among the Sioux seemed particularly interesting to Hennepin, who learned the Sioux language and recorded their customs. He baptized a small Indian child and tried rather unsuccessfully to promote Christianity. He recorded that he came into contact with other tribes, including the Assiniboines, who used buffalo dung as firewood.

Still, being a captive had its drawbacks, namely the lack of freedom. Life around the camp got considerably easier, especially since Aquipaguetin had adopted Hennepin—a welcome change from the chief's frequent attempts to kill him.

One day Hennepin saw his opportunity for freedom when the chiefs and warriors left for a buffalo hunt. Hennepin and one of his men said they were going for supplies which La Salle had promised to leave at the mouth of the Wisconsin River. So the men floated downstream toward what must have felt like freedom, but that was not the case.

After they had traveled 160 miles, Aquipaguetin appeared suddenly with ten warriors. The chief demanded to know if Hennepin had found the supplies that La Salle should have left for him. When Hennepin responded that they had not, Aquipaguetin himself went to the Wisconsin River, but found nothing.

Retracing their steps, Aquipaguetin and his warriors tracked Hennepin, this time finding him alone, after his traveling companion, Antoine Augelle Picard du Gay, had left to go hunting.

Aquipaguetin encountered Hennepin in a little hut he and du Gay had fashioned to keep the summer sun from pounding on them. Like before, Aquipaguetin seemed to appear out of nowhere, this time with a tomahawk in his hand. Hennepin drew two pocket pistols which du Gay had smuggled from the Sioux.

"Not intending to kill this would-be Indian father of mine, but only to frighten him and prevent his crushing me, in case he had that intention," Hennepin wrote.

Aquipaguetin didn't try to kill Hennepin. He only reprimanded him for traveling on the wrong bank of the river, where Hennepin would have been exposed to hostile tribes.

Any fight that may have remained in Hennepin disappeared when Aquipaguetin suggested that Hennepin come back to the village with him. History repeated itself, and Hennepin was again a captive of the Sioux.

As the Sioux and Hennepin were making their way back to the village at Mille Lacs, they encountered French explorer Daniel Greysolon Dulhut (or "Duluth" as we now know him).

Dulhut had heard from other explorers about the three Frenchmen being held by the Sioux, and he had been searching for Hennepin and his men. Dulhut had dealt with the Sioux in previous years, awing them with tales of Louis XIV's power. His storytelling had amazed the Sioux so much that they rushed to sign a peace treaty with him, fearing the wrath of the powerful French king.

When Dulhut met Hennepin, they formed an instant friendship. Both had served in the military in the battle of Seneffe, and Dulhut was outraged that the Sioux had stolen the priest's vestments. Rebuking the Sioux for the way they had treated Hennepin and his men and upset because they'd broken the peace treaty, Dulhut refused the peace calumet offered to him at Mille Lacs. He had come, he said, to take his French friends back to Canada, even going as far as telling the Sioux that Hennepin was his brother.

The Sioux chieftains feared Dulhut's wrath and willingly let Hennepin and his men go. It was September, and Hennepin had spent a little less than half a year as a captive.

Hennepin's life in the New World is well documented, but his death is not. The exact date and location are unknown. A single letter exists that suggests that he died in Rome sometime after 1701.

PIKE'S FORGOTTEN DISCOVERY

- 1806 -

JUST TWENTY-SIX YEARS OLD AND IN CHARGE of an exploratory party, Zebulon Pike stood by watching his dying men. They were fifteen hundred long, lonely miles from where they began, and a fall chill was in the air. The woods were thickly forested, and the Mississippi River was full of snags and back channels, making it a challenge to navigate.

Pike's right-hand man, Sergeant Kennerman was vomiting blood—and not just a little—but quarts of it. Another member of Pike's party was sick and suffering from a painful bladder infection.

Pike, a U.S. Army lieutenant who had been charged with finding the source of the Mississippi River, was driving his men to death.

"If I had no regard for my own health and constitution, I should have some for those poor fellows who were killing themselves to obey my orders," Pike recorded in his journal.

The twenty-two members of his expedition made camp somewhere near the present site of Little Falls, Minnesota, where they set to work unloading supplies and building a winter camp and stockade.

Most of his men would stay behind for the winter, resting with an ample supply of meat and whiskey. While wintering, the men fashioned small pine canoes for the detachment. Come spring these canoes would carry the men as they struck out to find the river's source.

However, Pike, the intrepid explorer, would push on during these colder months. He was determined to discover the Mississippi River's source. Pike was dedicated to his mission, to his country, and to his friend, General James Wilkinson, who had sent him on this journey.

Finally a few of the canoes were ready and Pike pushed on. He spent the day before his departure writing letters and leaving explicit orders with the men who would remain behind. He detailed instructions for the recuperating Kennerman, who would command the stockade while Pike pressed on.

Assuming everything would be fine at camp, Pike headed out. He took his two canoes to the head of the rapids; loaded his supplies, including ammunition; and then left them with a sentinel while he gave last-minute orders to the party.

Within one hour, the canoe carrying Pike's baggage and ammunition had sunk. The party tried salvaging some of the items, but most of them were lost, including the ammunition which was needed for survival and defense.

"The extent of the misfortune, the magnitude of which none can estimate, save only those in the same situation with ourselves, 1500 miles from civilized society; and in danger of losing the very means of defence, nay of existence," Pike recorded in his journal.

While the despair of the sunken canoe weighed heavy on him, he started the next morning by harvesting more timber for another canoe. He was convinced his canoe sank because it was too small, and he set about correcting the problem. As he tried drying the gunpowder from the previous day, it exploded, taking several tents along with it.

Pike sunk deeper into despair.

"I found myself powerfully attacked with the fantastics of the brain, called ennui, at the mention of which I had hitherto scoffed," Pike wrote. "But my books being packed up, I was like a person entranced and could easily conceive why so many persons who have been confined to remote places, acquired the habit of drinking to excess and many other vicious practices, which have been adopted merely to pass time."

More than a month had passed since his canoe sank and most of the gunpowder had literally gone up in smoke. He was eager to finish the mission, completely disregarding the hostile Minnesota winter and trusting the winter stockade to the men he later called a "dam'd set of Rascals."

Yet, by the time he set out, the river had frozen, making canoe travel impossible. Instead, Pike had two sleds constructed. The sleds could be pulled by soldiers standing two abreast. Six others were assigned to pull the canoes.

The harsh Minnesota winter was cruel and unforgiving, and Pike and his men often traveled less than five miles per day. Just four days after leaving the winter stockade another accident hindered Pike's expedition. One of the sleds had come too close to the river and fallen through the ice. Pike's men dove into the icy water to retrieve the equipment, but Pike's baggage, books, and cartridges, along with a little salvaged gunpowder were almost all destroyed.

The sled was soon repaired, and the party trudged on. Pike walked ahead of the crew, building fires, and selecting campsites along the way. He was oftentimes so cold and exhausted that he was unable to even write legibly in his journal.

Pike and his men were making steady but slow progress, and one morning Pike awoke to hear one of his men say, "God dam your souls, will you let the Lieutenant be burned to death?"

Fearing mutiny, his first instinct was to grab his guns, but he soon discovered his tent was on fire. Several tents were completely lost in the blaze. What few personal possessions he had not already lost—including his leggings, socks, moccasins, and bedding—had been charred. Pike's crew managed to salvage only three small containers of gunpowder, the difference between life and death, protection and vulnerability.

The march to the headwaters of the Mississippi would take nearly another month.

The winter was "remarkably cold," according to Pike. Even a little bit of whiskey remaining in a small keg congealed "to the consistency of honey." Earlier that day, Pike had spotted a large animal, which he said looked like a panther, but he dismissed it because he estimated the animal was at least twice as large as one seen on the lower Mississippi.

Eventually Pike reached the shores of Leech Lake, which he called Lake La Sang Sue. "I will not attempt to describe my feelings, on the accomplishments of my voyage, for this is the main source of the Mississippi," Pike recorded.

That night Pike would make it to the Northwest Company's post, where the men would be treated to "a good dish of coffee, biscuits, butter and cheese for supper."

Nearly two weeks later, while staying at the Northwest Company's post, Pike journeyed to Cass Lake, determining that it was "the upper source of the Mississippi."

Satisfied that he found the source of the river, the crew prepared for the journey back to the winter camp. While the travel to the headwaters had been dangerous, the return trip was no less exhausting. Pike stopped briefly because his feet had been cut and were bleeding from the racket strings of his snowshoes.

When the men reached the camp they had left nearly three

months earlier, Pike learned that Kennerman had used up most of the meat; drank or sold most of the whiskey; sold the flour, salt, and tobacco; and had even sold Pike's few remaining personal items.

Pike was furious and wrote in his journal, "Thus, after I had used, in going up the river with my party, the strictest economy, living upon two pounds of frozen venison a day; in order that we might have provision to carry us down in the spring; this fellow was squandering away the flour, pork, and liquor, during the winter, while we were starving with hunger and cold."

For his behavior Pike busted Kennerman down to the rank of private, a relatively mild punishment for jeopardizing the entire crew.

The expedition finally made it back to St. Louis on April 30, 1806, after having traveled over five thousand miles. Little fanfare awaited the group when they returned—General Wilkinson hadn't received presidential permission for the party before Pike set out to discover the Mississippi's source. Later, however, President Thomas Jefferson officially sanctioned the exploration.

Unfortunately, while Pike, dubbed by one biographer as "the poor man's Lewis and Clark," claimed to have found the headwaters of the Mississippi, he had actually failed. He had taken a wrong turn and merely found Leech Lake, a tributary of the Mississippi. Pike had not traveled far enough to discover the true source of the Mississippi River, Lake Itasca.

However, on his way to find the headwaters, Pike met with Sioux tribal leaders. Pike eventually signed a treaty purchasing a hundred thousand acres on both sides of the Mississippi River for a military post. The land included St. Anthony Falls. In exchange, Pike offered only some meager trade goods and two hundred dollars credit (of course, Pike didn't have cash on hand). The land that Pike purchased would later be home to Fort Snelling and the Twin Cities of Minneapolis and St. Paul.

PIPESTONE

On the Mountains of the Prairie,
On the great Red Pipe-stone Quarry,
Gitche Manito, the Mighty,
He the Master of Life, descending,
On the red crags of the Quarry . . .

—HENRY WADSWORTH LONGFELLOW,
"THE SONG OF HIAWATHA"

PIPESTONE OWES A LOT TO LONGFELLOW.

When Henry Wadsworth Longfellow penned "The Song of Hiawatha," he captured the imagination of a country with his words. From the actual accounts of the red rock quarries of Pipestone grew the mythical landscape of Pipestone. While Longfellow may be responsible for putting Pipestone in the public consciousness, the first white person to visit the site is something the public may never know.

For many years the title of "first white man" to discover Pipestone went to famous artist George Catlin. He earned the title the old-fashioned way—he gave it to himself. Making up a good story had always come natural to Catlin. He'd been a connoisseur of good stories since he was a young boy, and his mother told him about being taken captive by American Indians.

His mother's story is true. Catlin's mother and her family were taken captive during the American Revolutionary War when she was just seven. The story was never far from his mind, even during adulthood. His home state of Pennsylvania was still a frontier, and American cities were visited by American Indians frequently.

In 1823 Catlin witnessed a delegation of Native chieftains visiting Philadelphia. It was that event that drove Catlin out of his office as a lawyer and into the wilderness of America. Catlin's life would be pulled by two opposite forces—the world of the European settlers where he'd encounter fame and the world of the Native American tribes where he'd encounter amazing new cultures and artifacts.

From the start, Catlin had a vision: Rescue "a vanishing race" of people whose "looks and customs" were beginning to fade.

His journey to Pipestone began when he met up with General William Clark in St. Louis. Clark showed him a collection of pipes that were made of the red stone found in the quarries. Clark was famous because of his journey with Meriwether Lewis. Though the Lewis and Clark expedition never reached Pipestone, they traded for pipes and artifacts and heard many reliable accounts of the quarries from American Indians they encountered.

In September 1836, Catlin reached the Pipestone quarry area. Catlin discovered that the red rock was being quarried by different tribes. Instead of warring with each other for quarrying rights, the different tribes were at peace.

Catlin discovered why the land was so special when he heard a legend told by the Sioux. The legend told of the Great Spirit standing on a wall in the form of a large bird, calling the tribes around him. The Great Spirit then broke a piece of the red stone, made a pipe out of it, and smoked it. As smoke rolled over the tribes, he told his "red children" that the pipestone was their flesh, and they were made from it. The Great Spirit told the tribes that they must smoke to him through pipes made of pipestone, and the sacred stone must only be used for pipes. Since the flesh of all the tribes was the same, the land around it must be open to all the tribes—in other words, a neutral site. The Great Spirit forbade any weapons to be used on the land or even brought onto it.

Stories like these made Catlin famous when he returned and showed his drawings of "savages." Explorers who came after Catlin questioned his accounts of the legends he recorded. His portrayal of the land and the people wasn't completely inaccurate, but usually embellished and romanticized. It still provided, however, an original depiction of Native American life before settlers ventured into the area.

In addition to drawings and ideas, Catlin also brought back artifacts from his travels, including a sample of the quarried red rock. When he submitted it to geologist Charles Thomas Jackson in Boston, Jackson classified it as a new rock and promptly named it after Catlin. The rock quarried in Pipestone today is still called "catlinite."

Historians and archaeologists soon discovered many red stone in pipes from tribes across America made from catlinite, proving Americans were just learning what American Indians had known about for some time. Catlin proclaimed himself the "first white man" to discover Pipestone. Historian Sally Southwick also notes that Catlin, on his speaking tours, would discuss tribal lore about the Pipestone quarries. These tours, along with a lack of other credible sources, helped him rise to prominence.

But, back on the windswept prairies of Minnesota, a rugged trader was secretly seething.

"Catlin claims to be the first man that visited the pipe-stone, but this is not so. In 1830, I found a 6 lb. cannon-ball there," wrote Philander Prescott in his journal.

He would continue to write, harboring his grudge.

Prescott had traveled through the area several times, including the time he found a cannon ball there. Although he didn't record how it got there, Prescott said that it was used to break the red stone named for Catlin.

In 1831, half a decade before Catlin visited the site, Prescott recorded in a journal, "We got out a considerable quantity [of pipe-stone] but a good deal of it was shaly and full of seams, so we got only about 20 good pipes after working all day."

Even though Prescott claimed to have visited the quarries as early as 1830, even he may not have been the first white man to visit the site.

Both Catlin and Prescott acknowledged that Joseph Laframboise, a local trader, beat them to the site. Yet they dismissed Laframboise because he was of European and American Indian heritage.

Despite his critics, Catlin would leave an enduring gift to Pipestone. He embraced the idea of creating a park to preserve the American Indian culture there. In 1836 Catlin imagined a park where the "the world could see for ages to come the native Indian in his classic attire." It took 101 years for Catlin's idea to become a reality. In 1937 President Franklin D. Roosevelt signed the legislation that created part of what Catlin had dreamed. Today, Pipestone is preserved as a national monument.

Much like the controversy over who discovered Pipestone first, two Native-American groups remain locked in controversy over quarrying the stone and the sale of stone pipes.

Even though Pipestone is now a national monument, American Indians still quarry rock on the site, and the Yankton Sioux, the Lakota Sioux, and the National Congress of American Indians advocate that quarrying should only be done in special circumstances and then only for the purpose of sacred pipes. They also say that pipes should not be sold under any circumstances.

The Pipestone Native American community, however, an unrecognized band of Native people mostly of Dakota bands, argues that the Yankton Sioux do not have exclusive rights to the area and therefore cannot dictate how the pipes are to be used.

INSTANT CITY: JUST ADD RESIDENTS

- 1852 -

ON THE UNFORGIVING, GARBAGE-LINED STREETS OF New York City, back-broke workers trudged home, most too tired to even dream about a different life. Oily, grimy, and covered in coal dust, they were beaten down by the steam-filled factory halls, half deaf from the noise. The clean air and green forests of a land called "Minnesota" wasn't even on their minds.

But a few—the ones who were lucky enough to read—momentarily stopped to read Horace Greeley's *New York Tribune* and his admonition, "[G]o west, before you are fitted for no life but that of the factory."

Journeyman printer William Haddock was inspired by the promise of the American West, where gold seemed to be plentiful, and land was literally for the taking. The ghastly overcrowding of New York City made him yearn for a better life, and he couldn't help but take the advice of Greeley, a fellow printer.

Haddock printed some handbills for anyone interested in establishing "The Western Farm and Village Association." The purpose was "to settle on our Government's Lands in the West in such a manner that farmers, mechanics, manufacturers, and merchants may possess the advantages of the first purchase of the land, without paying the higher prices which a crowded population creates, and at the same time enjoy all the advantages of an intelligent and industrious community."

There were only two problems. First, the group had no land, nor maps. Second, few, if any of them, were farmers or manufacturers, and few were particularly well educated. They simply wanted to escape the oppressive city.

The group set about solving its larger problem: It had no land. The association hired Ransom Smith, an appropriately named explorer, who charged the association one dollar per day, plus expenses while he scouted out possible town sites. After two months of exploring Wisconsin and Iowa, he quit, causing panic among the hopeful New Yorkers.

Haddock and another member, Arthur Murphy took matters into their own inexperienced hands, setting off to succeed where Smith had failed. They set out for the West, arriving in La Crosse, Wisconsin, on February 26, 1852, just across the river from what would soon be Minnesota. They were eager to stake a claim and map a site. The Mississippi River was frozen, but they pushed on with skates, a buffalo robe, and some camp supplies. They hugged the Minnesota bank closely. So excited to be in pursuit of a new home, they barely stopped long enough to have lunch with an area trader, who must have chuckled at the two New York City men on skates. They passed what later would become Winona, dismissing it for a site because it seemed prone to flooding. The two New Yorkers laughed off Winona because they recalled only four or five "shanties" there at the time and "already . . . the claimants [have] begun to fight about their claims."

The two skaters continued traveling up what they believed to be the Mississippi River. Instead, they were skating up a large slough of the river, later named Straight Slough. There, they discovered Rollingstone Valley, a perfect place, they thought, for their new home.

"Indeed, I may say that it is beautiful, and throws [Winona] and Prairie La Crosse entirely in the shade," Haddock wrote boastfully in his journal.

On March 31, members back in New York City drew lots for parcels of land in their new hometown. All eligible members' names were placed in one hat, the lot numbers in another. They drew until everyone had one parcel of land. One week later, an advance party of eleven boarded a train headed west. The rest of the group followed a week later.

For 50 cents a head, they traveled by steamboat as far as Wapasha's Prairie, now the site of Winona. There they unloaded their meager possessions and embarked to Minnesota City. Travel became much more difficult at that point. As if trains and steamboats weren't bad enough, the men struggled to get their possessions and supplies through the muddy sloughs and marshes. The few other settlers who lived in Winona couldn't be concerned with the new naive Yankees because they, too, were fighting for survival on the rugged frontier of early Minnesota.

When the pioneers finally reached the site, they busied themselves building "gopher huts," shelters of timber built over a hole about a foot deep to ensure head room. The primitive confines were a far cry from the bustling streets and tenements of New York City. This was much different than the utopia they'd seen on printed maps of their settlement complete with glass greenhouses and city parks.

Two weeks later on May 1, more settlers began to arrive.

Women and children crowded into one settler's cabin, while men camped in the huts. A large tent was also constructed for a temporary shelter. Over five hundred men, women, and children had come over one thousand miles and were faced with the prospect of having to literally create a town from the ground up. One member of the party, E. M. Drew wrote in his journal, "They could do nothing because they had no lumber and it seems did nothing toward getting any. They sat around as though waiting for the lumber to come to them." The group survived on mush and molasses, Drew recalled.

Bad news and disease quickly broke out in the camp. It started when William Christie barely made it to camp. Christie managed to ford the engorged river. When he arrived in camp, he promptly complained of not feeling well. He didn't even change his wet clothes before lying down. Christie never got up and died just hours later of what the new Minnesotans believed to be cholera.

E. H. Johnson built Christie's coffin from unseasoned pine boards which were lying on the bank of the river. A short funeral service was held in the open air in front of the shanty where he died. His death was the first one noted in the county, but before summer was over, death would visit frequently.

To make matters worse, the settlers began to realize that the water of the Straight Slough, the tributary branch of the Mississippi, was simply not deep enough for steamboats. The newly minted settlers insisted the slough was indeed passable for steamboats. It *had to be,* because it was the only connection the fledgling camp had to the rest of the world.

The steamboat the *Nominee* attempted to navigate the slough, but failed, getting stuck on a sandbar, just as its captain, Orrin Smith, had predicted. It was actually happy news for Smith, who breathed a sigh of relief. If steamboats couldn't reach Minnesota City, it wouldn't become a rival to his recently founded community, Montezuma (the name of Winona until 1853). Minnesota City pioneers had founded their city six miles from the nearest Mississippi River landing.

Settlers wanting to go to Minnesota City first had to land at Montezuma, sometimes staying days before they could get transportation to Minnesota City. They often had to leave goods on the riverbanks for over a week as they waited. When the settlers finally struck out for Minnesota City, it often meant fording Straight Slough. Men were forced to swim with the cattle, taking off their clothes, and guiding the rafts. Sometimes a wood boat, called a *Macedonian* helped members transport goods and passengers, but it was not always available.

Summer arrived and it was hot and sticky. People began getting sick. The settlers blamed the weather for creating the sickness. The little piece of paradise had quickly turned into a hell. Minnesota City didn't have a doctor or reliable transportation, and the camp began to lose hope. Even Haddock's wife, who arrived on July 20, died on August 24.

Association members started to leave, including Haddock himself. Haddock left the settlement to find the printing press he had ordered from Dubuque, Iowa. He had begun soliciting subscriptions for a paper he called the *Minnesota City Standard*. That paper never made it to the first edition. After a brief stop in Dubuque, Haddock relocated to Anamosa, Iowa, beginning a paper there.

Meanwhile, sickness continued to be problem in Minnesota City. After Robert Taylor lost two children, he moved his ailing wife to La Crosse. It didn't help. Shortly after the move, she died.

Most historians agree cholera was the culprit, and many theorize that the settlers brought the disease with them.

Still, other settlers toughed out the primitive new environment during the winter. The population, however, continued to decrease due to disease and desertion. Really, they were just fulfilling Greeley's prophecy. Few had noticed that Greeley, who had taken an interest in their cause, had predicted it would fail because of a "defective plan." He had sensed the lithographed image of a beautiful prairie and a twenty-four-acre park with a glass winter house where citizens could grow vegetables was little more than a fanciful embellishment.

As soon as spring arrived and the river thawed, more settlers fled downriver and across land to more friendly locales. The population, which had briefly established the largest colony in Minnesota, was crumbling. Over five hundred was quickly reduced to less than three hundred.

Minnesota City would never become the utopia or bustling community its architects had hoped. However, it could boast the first church and first school in Winona County. Just three years later, an informal census revealed that half of the 350 people living in the valley were original members of the Western Farm and Village Association. That means approximately 225 others had left or died. It is an unpleasant reminder that the American West was a hostile, unforgiving, and treacherous land, far from the paradise eastern books and newspapers had described.

ROBBING ST. PETER TO PAY ST. PAUL

- 1857 -

THE SUN BARELY SNUCK IN FROM UNDER a pulled window shade. The haze from cigar smoke cast an eerie light on the poker table. And for one guiltless legislator, the making of laws was merely a bump in his daily routine of poker, booze, and cigars.

While Joseph "Jolly Joe" Rolette's fellow lawmakers were in session, he sat drinking some of the finest wine that could be imported to the territorial capital city. And after wine, he switched to whiskey at the Fuller House, a local lodge where the buckskin-clad legislator sat in hiding. Rolette's self-imposed exile from the 1857 legislature was more like a vacation.

A few floors down, locked in Truman H. Smith's vault, was a bill from the legislature that would have moved the state capitol from St. Paul to St. Peter, Minnesota. Rolette had simply taken the bill, just before it was needed, and walked out of the legislature.

It's easy to see how he escaped. The legislators were busy at work, focusing on a different, but important matter: They were preparing for what would be a successful run at statehood.

During that legislative session, lawmakers wrestled with where to put the capitol when Minnesota entered the union. While debates centered around how to draw the state's boundary lines (whether it should be short and wide, or as it currently is, not as wide, but extending to the border of Canada), a powerful group of legislators advocated moving the capitol to St. Peter, a location in the southern region. The bill to have the capitol moved was introduced by a politician who was said to have strong ties to the Saint Peter Company, which owned the land for the newly proposed capitol.

Meanwhile, territorial governor Willis Gorman, who had been president of the Saint Peter Company until 1856, was poised to give his approval to the measure. Although many saw the move for what it was—a political power grab—the idea was not without merit.

Many felt the state's boundaries would be drawn differently. That is, the northern part of the territory would become part of another state or territory. Moreover, development, farming, and railroad interests had grown strong in the southern and central part of the state in places like Winona and Mankato.

Even though Rolette had come from Pembina, a place that would later become part of North Dakota, not Minnesota, he and his friends favored leaving the capitol right where it was. Sensing the bill was all but guaranteed to pass and gain the governor's approval, Rolette used his position and in one extraordinary show of brazenness, sunk the plans for moving the capitol.

Rolette played the ultimate political trump card: As the chairman of the enrollment committee, he was in charge of producing final versions of legislative bills. His earlier attempts to hold up the bill in committee had failed after St. Peter advocates forced a motion to require the committee to produce the bill immediately. By the time the motion made it through, Rolette had disappeared with the bill in hand. If there was no bill, it could not be read or passed.

Rolette hid at the Fuller House for eight days. But, there was only one slight hiccup: According to legislative rules, once the call for the bill had been issued, it could not be suspended unless by a two-thirds vote. Because neither side had a supermajority, the committee couldn't even move on to other work. Other committee business could not proceed without unanimous consent, another impossibility since Rolette was absent without leave. The committee couldn't adjourn either. To solve that, cots were brought in. Meals were also brought in as the committee bivouacked in the committee chambers.

For 123 hours they waited, falling just one vote shy of suspending the report.

St. Peter advocates, determined not to be defeated by few procedural technicalities, mustered enough support to order another copy of the bill prepared. However, the remaining committee explained there had been enough changes made to the bill that the only accurate copy was the one Rolette had with him. Rolette not only was holding up the capitol bill, but because business was stalled, the House amendments to the bill for a constitutional convention were also delayed, risking more than just a capitol.

According to the law, the territorial legislature was only authorized to meet for sixty days. At midnight on Saturday, March 7, 1857, the session would come to an end, with or without Rolette.

That day approached quickly. Just seconds before midnight as the presiding officer had his gavel raised, Rolette appeared.

Only enough time remained for Rolette to say, "Mr. Chairman," before the gavel struck, signaling the session's end. Even the supporters of a St. Peter capitol couldn't help chuckling at Rolette's outrageous antics. The following day, he was given a torchlight parade through the streets of a grateful St. Paul.

Supporters of St. Peter, however, weren't ready to give up.

They had made another copy of the bill. When the committee refused to enroll it, supporters short-circuited the legislative process, taking the bill to Governor Gorman who signed the bill. To give the illegally passed bill more weight, Gorman also ordered three commissioners to set up a suitable new building for the capitol. Without funds or a proper vote, those commissioners bought a tract of land in St. Peter. A frame for the new capitol was constructed at a cost of five thousand dollars.

Almost as soon as that happened, St. Paul supporters filed a lawsuit challenging the governor.

Judge Rensselaer Nelson quickly ruled that the legislature had not passed the law, and that the capitol would remain in St. Paul.

The new structure, "the Capitol," instead became the Nicollet County Courthouse, a reminder of robbing St. Paul and not paying St. Peter.

For decades, Minnesota history textbooks taught students that Rolette had absconded with the bill. However, he had merely bent the rules of order to his advantage. As the chairman of the committee, he was certainly entitled to handle the bill. Even fewer textbooks mention that Rolette—a larger-than-life figure in early Minnesota history—would die a pauper less than fifteen years later.

DAKOTA SIOUX WAR

- 1862 -

WHEN FOUR MDEWAKANTON SIOUX ARRIVED at Howard Baker's Acton farm, he was glad for the company on the lonely Minnesota prairie. The four Sioux challenged Baker and his neighbor, Robinson Jones, to a shooting contest—guns and targets being about the only two things not in short supply. They quickly agreed to a friendly competition of marksmanship, smiling and picking out the targets. Little did the two neighbors know that they would be the targets, and the game was a trap.

The two neighbors fired at the target first. Then, the Sioux took aim. After shooting at the target, the Sioux immediately reloaded and turned their guns on Jones and Baker, slaughtering them. The four warriors, Killing Ghost, Breaking Up, Runs Against Something When Crawling, and Brown Wing, scurried back to tell their leader, Red Middle Voice, what had happened.

Congress had been late sending to the Sioux the yearly annuity payments guaranteed by treaty. A crop failure made the situation worse, and the Sioux were being pushed farther and farther westward by the waves of land-hungry immigrants. The new expansion also brought traders, cities, and railroads. The American Indians, especially the Sioux, had been pushed to an almost untenably small reservation. The tribe depended on hunting large tracts of land for its survival. Once the government stripped their hunting ground from them, they became reliant upon government annuities. To make matters worse, unscrupulous traders preyed on the hungry Sioux, charging high prices and plying tribal members with alcohol.

At first Sioux chief Little Crow was skeptical of waging war on the settlers and government. "The white men are like locusts when they fly so thick that the sky is a snowstorm," he told the warriors before the Sioux opted for war. "Yes, they fight among themselves, but if you strike one of them, they will all turn upon you and devour you and your women and little children, just as the locusts in their time fall on the trees and devour all the leaves in one day."

Still, the band of young warriors had already attacked and Little Crow knew the massacre of the Acton neighbors would not go unpunished. So, rather than wait for the inevitable swarm of white locusts, the Sioux went on the warpath.

On August 18 over two hundred warriors flooded into the Lower Sioux Agency, where they killed traders and clerks and then began looting. The war party was so intent on looting that it didn't notice the fifty residents hurriedly escaping to Fort Ridgely, fifteen miles away.

Meanwhile, the Sioux continued their attack on the countryside, killing men; taking women, children, and livestock; and burning buildings as they went. More than twenty soldiers died when Captain John Marsh marched his troops from Fort Ridgely to engage the Sioux. The Sioux had anticipated a response from the fort and were waiting to ambush the troops.

The Minnesota troops had never fought Indians before and were taken by surprise. Residents of nearby counties were also shocked when hostile war parties arrived, killing farmers and settlers, many of whom had no advance warning and didn't suspect the impending danger. Almost no area newspapers told of the rising tensions between the traders and the Sioux in the days before the outbreak.

The first day of battle was a complete victory for the Sioux. By nightfall, Governor Alexander Ramsey began making arrangements for troops from Fort Snelling.

The Sioux warriors rightfully believed that if they decimated Fort Ridgely, it would clear the way to St. Paul, where they could eventually push all the settlers back to the east side of the Mississippi River. Chief Big Eagle said, "We thought the fort was the door to the valley as far as to St. Paul."

Two subsequent assaults on Fort Ridgely and New Ulm were thwarted, giving the Sioux warriors more of a fight than on the first day. White defenders of New Ulm burned nearly two hundred buildings in the town in order to defend it. Settlers had to largely rely on themselves for protection or escape.

Governor Ramsey authorized General Henry Sibley, a former trader and later governor of Minnesota, to march from Fort Snelling to push back the Sioux. Sibley and his forces didn't arrive until August 27, days after the attacks on New Ulm and Fort Ridgely.

One of the final major battles of the war was waged at Birch Coulee. A burial party of civilians and soldiers had been dispatched, but told to take caution. When the party came together that night, they were not alone. The Sioux once again lay in ambush. Two hundred Sioux warriors, some wearing turbans of grass, were waiting and opened fire. A fierce battle ensued for over twenty-four hours, making it one of the bloodiest and most prolonged battles of the war. Soldiers sought protection behind dead horses and attempted to dig in using the same spades they had used to bury fellow fallen comrades. Some were saved when Sibley's force heard the distinct noise of gunfire and came to the rescue.

The last campaign, the Battle of Wood Lake, took place September 23, 1862. Even though a few isolated skirmishes flared up after the battle, Wood Lake proved to the Sioux warriors that the locusts had come and the gateway to St. Paul was all but impossible.

After the Battle of Wood Lake, with the Sioux beaten, Sibley turned his attention to rescuing the countless settlers—mostly women and children—who were held in captivity. Newspapers and popular literature of the day were chock-full of accounts of settlers being taken prisoner by Indians and of the heroes who rescued them. Sibley wanted nothing but the full, unconditional release of the prisoners. Sibley wanted to be one of those heroes. Sensing that no good could come from further hostilities; the Sioux released 269 captives at a place called Camp Release.

The Dakota Sioux War was over, but not its aftermath, which sent panic westward. Angry settlers cried out for vengeance in Minnesota.

In all, almost six hundred people, including settlers, warriors, and soldiers lost their lives. Little did Chief Little Crow know that his words of advice right after the Acton murders would turn out to be prophetic.

"Kill one, two, ten and ten times ten and they will come and kill you," Little Crow said. "You are like dogs in the hot moon, when they snap at their own shadows. We are only little herds of buffaloes left scattered; the great herds that covered the prairies are no more."

THE LARGEST EXECUTION
IN U.S. HISTORY

- 1862 -

THE GROUP OF SIOUX WARRIORS SAT TOGETHER, talking. Some family members came to see them. To ready themselves, they painted their faces in bold, bright colors—aquamarine and vermillion—the colors of warriors. They urged family members to be strong. The thick, sweet smoke of pipes filled the air. Some talked of eternal happiness and the Great Spirit. Some friends cried. Father Augustin Ravoux spent Christmas Day night preaching to them, baptizing and administering communion to a few new converts.

The group of Sioux warriors were not preparing for a holiday celebration or battle. These were their final hours. The thirty-eight warriors were preparing for their own executions. They had been sentenced to hang for their part in the Dakota Sioux War.

One by one, each warrior had a rope placed around his neck. The drum beat started. In a final sign of solidarity, the thirty-eight warriors clasped hands. On the third drum beat, the scaffolding fell. Thirty-seven bodies went limp.

One, Rattling Runner, dropped to the ground when his rope broke. The authorities clumsily rushed over and rehanged the warrior. According to the accounts, he was already dead. A small cheer went up from those watching the spectacle that had begun months earlier at an Acton homestead. The wailing of the Indian women, who were being kept close by, could be heard. Twenty minutes later four wagons took the bodies to be buried, completing the largest execution in U.S. history.

With more than a little vengeance, the military, led by General Henry Sibley condemned 303 Sioux warriors for their supposed roles in the six-week conflict. Some historians believe the Dakota Sioux War resulted in as many as a thousand casualties, although six hundred seems to be a more probable number. The residents of the state were ready to exact their pound of flesh. Sibley and Governor Alexander Ramsey saw the six-week conflict not as one of an almost endless number of Indian battles and wars happening in the West, but as a massacre that needed avenging. Sibley couldn't help but see the anger almost daily in the newspapers. One rag cried: "Death to the Barbarians."

Over four hundred Sioux warriors were prisoners at Camp Release, a makeshift camp. Sibley started the trials there. Some recently freed Sioux stuck around to testify against the prisoners. Despite Sibley's promise to only prosecute those who were guilty of bloodshed, he did not concern himself with due process and fair representation. Only one person, Reverend Stephen Riggs, served as an investigator. His job was mainly to listen to the testimony of pioneer eyewitnesses and other "cut hairs" (Indians who had adopted a European lifestyle and who had been friendly to the whites) and then bring charges against those being held. A few Sioux were used to testify multiple times against other Sioux tribe members. Many Indians were convicted on nothing more than memories and hearsay.

Lieutenant Rollin Olin served as the judge advocate even though he never had any legal training and was a mere twenty-two years old. He was assisted by one lawyer. No accused Sioux were given any measure of due process and not one attorney was provided. Many were convicted based on their own testimony, which was offered in broken English or was translated. Some cases were tried in less than five minutes.

Sibley urged the kangaroo court to abandon formalities and niceties so that justice could be swiftly administered to appease the fury of an angered state. Over forty cases were heard in one day alone.

Almost four hundred had been tried by November 5, 1862. Sibley wanted the 303 death sentences to be carried out right away, but waited for approval from his superior, Major General John Pope, who waited for approval from the president. In 1862 the country was also embroiled in the Civil War and President Abraham Lincoln insisted on reviewing the case of every condemned person. Pope had forwarded the cases, all 303, to Lincoln.

Pope and Sibley expected the verdicts to be quickly confirmed, but the president was concerned with the prospect of "lynching, within the forms of martial law."

Lincoln himself had been raised on stories of Indian atrocities as his own grandfather had been killed by Indians in Kentucky. Still, Lincoln insisted there was a difference between those who killed, raped, and looted indiscriminately in a massacre and those who fought in war. As word of Lincoln's deliberation spread, the calls for mob violence grew.

As the condemned warriors were moved from New Ulm to Mankato, a mob pelted the prisoners with bricks, killing two Indians. The guards arrested a dozen people just to ensure their cargo got to Mankato.

Minnesota's congressional delegation pleaded with Lincoln for immediate action. Governor Ramsey telegraphed Lincoln warning, "Nothing but the speedy execution of the tried and convicted will save us from the scenes of outrage."

Lincoln delayed his decision until December 6, 1862, when he reduced the number of condemned from 303 to thirty-eight.

Colonel Stephan Miller delivered Lincoln's news and gave the message to an interpreter: "[T]heir Great Father at Washington, after carefully reading what the witnesses testified to in their several trials, has come to the conclusion that they have each been guilty of wantonly and wickedly murdering his white children."

As the sentence was being read, few warriors showed emotion. A couple smoked a pipe. The message was partially correct. Lincoln had spent hours poring over the files, and, even though he reduced the number of those going to the gallows, the testimony had been conjecture with little evidence.

Little Crow, the Sioux leader who reluctantly agreed to make war on the settlers, had escaped death, but not for long. He had run away after the battles, but returned the next year. On July 3, 1863, Nathan Lamson caught Little Crow picking raspberries outside of Hutchinson, Minnesota, and killed him. It provided Hutchinson residents with a perfect Fourth of July celebration as they put firecrackers in the corpse's ears and nostrils. Little Crow's body was thrown in a garbage pit, but his head was kept by Lamson so that he could turn it in to collect the bounty that had been placed on Little Crow's head by Major General Pope.

Episcopal bishop Henry Whipple had traveled to Washington, D.C., to beg Lincoln for mercy on the condemned prisoners, telling him the federal corruption in the Indian Affairs Department was as much to blame for the recent bloodshed as the Sioux themselves. When Lincoln met with Whipple, he said he felt the enormity of the Indian situation "down to [his] boots." And, it would be Whipple's prophecy that would become true both in the Civil War and when dealing with American Indians: "A nation which sowed robbery would reap a harvest of blood."

THE FOUNDING OF THE MAYO CLINIC

- 1865 -

DR. WILLIAM W. MAYO LED A RESTLESS LIFE. Being an itinerant doctor in Indiana just didn't seem to cure this prairie doctor's wanderlust. He had traveled across the world to become doctor, and he'd travel a little more to find happiness.

His first stop in Minnesota was in St. Paul, where he quickly discovered that the city had too many doctors. He soon found out that the little burg of Le Sueur was in need of a doctor and traveled there, where he quickly became one of the village's leading luminaries.

But the quiet doctor surprised everyone when the Dakota Sioux attacked New Ulm, a town not far from Le Sueur. Mayo quickly rushed to help the wounded and set up a makeshift hospital in a boardinghouse. As the wounded started to arrive and as the men standing guard heard gunshots and war cries, they lost heart and ran to take cover with the women and children in cellars.

Mayo attended to the wounded, but soon realized the guards outside the doors were leaving. He chased after them incensed, picked up pitchforks, and shoved them into the hands of the cowardly men. When one man asked what they should do if the marauders approached, Mayo curtly replied, "Run your forks through them, of course."

The battle raged on, moving closer and closer to the makeshift hospital. Fifty men were wounded in the first hour and a half of fighting. Men helping Mayo ripped the doors off the hinges and used them as stretchers. When Mayo again spied several men sneaking away from guard duty as he was performing an amputation, he came barreling after them with a bloody knife.

It might be said that the first Mayo Clinic was established during the Dakota Sioux War in New Ulm. The same dedication to patient care is still evident today at a much larger clinic in Rochester sans pitchforks and a knife-wielding immigrant doctor.

After the Dakota Sioux War ended, the U.S. government hanged thirty-eight Sioux warriors in Mankato. Mayo and his medical colleagues exhumed the bodies of the hanged men. Cadavers were in short supply, especially out in the remote Minnesota wilderness. They took the bodies in the name of science, and Mayo claimed the body of Cut Nose, a fierce Sioux warrior. He had the skeleton cleaned and articulated. Cut Nose's skeleton provided Mayo's two young sons, William and Charlie, some of their first anatomy lessons.

By this time, Mayo had earned quite a reputation as a courageous doctor and a loyal Republican. Mayo's reputation earned him a political appointment as the examining surgeon for Minnesota's First Congressional District during the Civil War, which meant he uprooted his family and moved from Le Sueur to Rochester, Minnesota.

Mayo's new position required him to examine all the Civil War draftees. Men would go to extreme measures to escape being drafted into the Union army. Stories of men cutting off their own fingers abound, and Mayo developed a reputation for sniffing out frauds.

For example, men claimed their arms or legs were permanently rigid, but Mayo discovered a whiff of chloroform would reveal if such a disability was real or fake.

Enrollment in the army was a tricky business—lawyers and some very shady characters often lurked right outside the examiner's office, promising to get scared men out of going to war. Bribery and jealousy were rampant. By the end of February 1865, Mayo was being investigated for examining men after hours. Though not technically against the law, Mayo's practice of seeing possible draftees after hours gave rise to suspicions that he was writing too many exemptions from military service. Mayo argued he was only seeing patients after hours as a convenience.

The army found little to quibble with when reviewing his list of exemptions. In fact, 105 of 153 exemptions fit a "strict" construction of regulations. Yet, the army was also concerned about the appearance of bribery and dismissed Mayo from military service for "receiving fees for private examinations."

Seemingly undeterred, Mayo bought a small piece of ground on Third Street in downtown Rochester in the summer of 1865. He decorated the room, a primitive clinic, with a plaster bust of President Abraham Lincoln. A simple, but effective newspaper advertisement read:

DR. MAYO
Office on Third Street
Rochester, Minn.

By the end of the year, Mayo was not only in private practice, he was also leading an effort to build a library. Soon he was leading a campaign for a Rochester school that would become the talk of the state. It would be a brick, five-story, sixteen-room building with furnace heat and a price tag of sixty-five thousand dollars.

Almost two decades would pass as Mayo's reputation grew increasingly prominent, helped in part by his two brilliant sons, Charlie and William, both of whom had followed in their father's footsteps and became doctors.

Before their deaths, the Mayo brothers turned over their entire fortune and medical practice to a foundation that would continue to run a clinic for the health of all who walked through its doors. Funds were set up to give back to communities and to a medical school that would produce the next generation of doctors. The Mayo Clinic also has a history that is quite unlike any other clinic. When the brothers created the Mayo Foundation, most lauded the decision, including the publication, *Commerce and Finance,* which gushed:

> *Of all the wonder stories of America, there is hardly one to surpass that of the Mayo brothers . . . What an inspiration their lives must be! Two plain American boys who made an obscure village a shrine of hope to suffering humanity; who have been the means of saving thousands of lives and averting untold pain and anguish; and who crowned their careers by giving their all for the benefits of mankind.*

WATKINS AT YOUR FRONT DOOR

- 1868 -

IN 1868 PATENT MEDICINES WERE ALL THE RAGE. Tonics, elixirs, compounds, pills, potions, and syrups promised miracles but often gave more hope than help. Most patent medicines could be grouped in two categories, useless or downright dangerous.

But, whatever the ailment, a pioneer druggist had a remedy. A young J. R. Watkins secured the right to manufacture Dr. Ward's Liniment, a concoction touted as a veritable panacea. Richard Ward was a Cincinnati doctor who had formulated the compound of camphor, extract of hot red peppers, oil of spruce, and other natural ingredients in an alcohol base.

In Plainview, Minnesota, where he lived, Watkins took the recipe and mixed up the cure-all in his own kitchen and then bottled it by hand in a woodshed. Then, Watkins set out in a horse-and-buggy carriage to peddle his potion. The liniment was used to cure anything from constipation to the achy, sore muscles of the farmers working the southern Minnesota farmland.

Watkins's fame as a salesman spread through the pioneer country and his liniment gained a reputation for being effective, which was remarkable in a golden age of medicinal quackery.

J. R. Watkins's salesmanship was outstanding. Generations of advertisers and manufacturers would be forever indebted to Watkins as the grandfather of the "money back guarantee."

As a marketing ploy, Watkins boasted, "satisfaction guaranteed or money back." He had his liniment bottles made with a trial mark, a notch or mark on the side of the bottle. The idea was simple: The customer could use the liniment down to the trial mark; if the customer wasn't satisfied, they could return the remainder of the bottle to Watkins the next time his horse and buggy passed by. Watkins would refund the money, but very few appear to have ever been returned. Today the idea seems common place, but it was novel in the day of buyer beware.

In time Watkins assembled a small, but growing army of other salesmen in southern Minnesota. They quickly became known to their customers as the "Watkins man." Word of Watkins products had begun to spread. As a "wagon salesman," Watkins embraced the idea of running a whole fleet of salesmen and saw new markets all over the country, many of them in remote locales not unlike Plainview.

By 1940 the company would expand to include distribution warehouses across the country and Canada. The Watkins Company would grow to over two hundred products and include a popular line of cosmetics, Mary King, aimed at a new generation of 1920s women who had embraced once-taboo cosmetics. The Watkins Company also sold feed and farm products and even ventured into the automotive market, manufacturing spark plugs and tires for a time. The automotive line went bust very quickly, but the company briefly produced a tire which had tread formed to look like the letter *W,* for Watkins.

In 1940 the company boasted ten thousand salesmen across the world. Four years later, that number was fifteen thousand. The Watkins Company enjoyed the distinction of being called "Depression proof," by many business publications. Despite being Depression proof, Watkins wasn't change proof. The American lifestyle had changed rapidly by the 1960s. More women worked outside the home, and the market had become saturated with door-to-door salesmen selling encyclopedias, knives, and vacuums. Watkins entrepreneurial spirit had also caused the company to venture into making different lines of products, many of which were never successful. The company started to falter, and by the mid-1970s creditors had begun to call in loans.

Irwin Jacobs bought the bankrupt company on December 28, 1978, and began a slow, almost two-decade process of rebuilding the historic Minnesota company. Instead of going door to door, the company eventually went computer to computer, developing its business one customer at a time, just like Watkins did after mixing up his first batch of liniment in Plainview. Now called Watkins Incorporated, the company's business is still booming to this day.

FROZEN STIFF:
THE MINNESOTA BLIZZARD

- 1873 -

PER HANSA AND HIS FRIENDS WERE ALMOST GIDDY as they gathered firewood. A storm a few days earlier had stopped everything, and now he and his friends were enjoying the mild temperature. The sunshine and crisp air felt good, especially after being cooped up in the stuffiness of his home watching snow and ice blanket everything.

As he gathered firewood, Hansa noticed a dark gray cloud on the horizon. The gray cloud drew closer and closer, moving toward them as if it had eyes. As it seemed to rush on top of them, everything fell silent for a moment.

Then a massive squall of snow hit. The snow was so thick Hansa couldn't see his hand on his outstretched arm. The wind whipped and howled in all directions, snow began to fall, and the air hissed and mingled with the deep rumblings of thunder.

This is how the Blizzard of 1873 started for Hansa and so many others across the entire state. It crept up on Minnesota suddenly and immediately turned deadly.

For most of January 8, the weather had been mild. Temperatures had even risen high enough to melt some of the snow which had gathered a few days earlier. Folks ventured out, making trips for supplies and fresh air. But, between three and four that afternoon, the weather took a dramatic turn. More than one source said the temperature dropped ten degrees in a minute. The wind and cold weather brought a fine, driving snow that made it impossible to see one step ahead. The snow fell so quickly and the wind blew so hard that many said the sky was dark in a matter of a few minutes. The storm roared across the state, and by nightfall it held over a dozen counties in its grip.

"The flakes came down in rather close proximity on Tuesday," said the *Minneapolis Tribune,* an understatement, "and during the night, well, no words of ours can do the subject justice."

The entire state ground to a stormy, screeching halt for almost sixty hours as the blizzard covered the land of ten thousand lakes and the temperatures plunged below zero.

As thousands remained trapped in whatever shelter they could find, hundreds were trying to survive the storm. Others died the kind of deaths that are found in some third-rate novels.

In New Ulm, Minnesota, one boy who lived across the road from his schoolhouse never made it home. His body was found eight miles away from the schoolhouse when the storm subsided. The storm's fierce wind and snow completely disoriented anyone who ventured into it. One man died less than ten feet from his own house after managing to drive his team of horses back home.

Another man, who was traveling near Glyndon, Minnesota, became lost in the storm. He unhitched the team of oxen and left to find shelter with his trusty companion, a dog. He was found two days later, less than four hundred feet away from where he'd left his oxen, his dog by his side. One man, thinking the dog to be a wolf or a fox, shot the dog only to discover the frozen body.

A farmer in Willmar, Minnesota, where the storm seemed to be extremely intense, reported that on his way to safety, he passed no fewer than five oxen teams frozen to death.

There were plenty of too-close-for-comfort stories as well. One man luckily found shelter in a barn only to later realize he was just a few feet from the adjoining farmhouse.

In Elizabeth City, Minnesota, another man was found near the town wandering around. He had become snow blind and had stumbled around aimlessly trying to find help.

One woman went out to get clothes from the clothesline just as the storm began. The wind blew the clothing from her hand. She rushed to pick it up and by the time she caught it, she couldn't see the house. She stumbled in what she thought was the direction of her home. As her feet and hands grew numb, she literally ran into the side of the house. She then began to work her way along the house until she felt a door. Using what strength she had left, she forced the door open, surprising a couple of carpenters who had been building a house not far away from hers.

As one nameless newspaper correspondent later said, "for to wander was almost certain death."

Hundreds escaped death's frozen grip, but didn't manage to exit unscathed. Many people survived with frostbitten limbs that needed amputation. Farmers lost entire herds of livestock, which froze to death in the fields.

To try to make up for the devastating loss, the Minnesota legislature set up a relief fund. Ninety-four families in thirty-four counties received aid from the bill for their losses. The bill also contained funding for hospital care, medical services, and surgical operations, mainly amputation of frostbitten limbs.

Railroads also halted. Huge drifts of snow blocked many engines and trains. Even snow plows and crews of men had to wait days to get to the stranded passengers. Most passengers had heat and limited supplies. More importantly, they had shelter from the blinding snow and howling winds. One conductor, whose train was stranded near Kasot, Minnesota, spent time reading select poems on solitude to boys on the train. The account of this doesn't mention what the captive audience thought; however, the conductor wired for help, asking "the train dispatcher to inform other conductors that the snow drift has no flags out, and to be careful not to run into it."

One plow team working on the Winona–St. Peter Railroad near St. Peter, Minnesota, noticed a dark object about a hundred yards from the railroad station at the little town of Oshawa. When the crew checked on it, they discovered a team of horses and a cutter with two men wearing buffalo coats, sitting straight up, and frozen solid.

Not only had the storm come suddenly and ominously, it had also brought electricity with it.

"The air was so full of electricity today, that both the batteries [of the telegraph] were thrown off in the railroad office and the wire operated entirely from the electricity in the air, this primeval mother of all batteries furnishing a strong current for the business," the *Winona Republican* reported. "This electrical phenomena furnished considerable sport at the telegraph office in the depot. Before the battery was taken off, the wires were so highly charged that a continual popping and flashing was going on, creating no little alarm among some of the boys who don't 'hanker' after shocks."

On the first day of the storm, the road between Minneapolis and St. Paul, the Twin Cities, was blocked by snowdrifts that varied from five to ten feet deep in some places.

"A snow plow batters against them in vain," the *Minneapolis Tribune* reported.

Though trains and other transportation seemed to be stuck, it didn't stop the newspaper from trying to deliver the news. One of the *Minneapolis Tribune*'s reporters sent a messenger to attempt to get from St. Paul, where the legislature was in session, back to the newspaper office.

"I'll start a messenger now, on horseback, with report. If he does not arrive by 9, let me know," the correspondent wired.

The paper went on to say, "If our readers find a full Legislative report, they may understand it as a signal that our Knight of the Snowbanks has got through: If not, that his name is not Stanley and he has succumbed."

Discovery of the dead—and even those that had barely survived—took days. People worked nonstop to get out from under the snow. Even recovery of the dead was slowed by temperatures which plummeted after the blizzard subsided. On January 17, 1873, the *Minneapolis Tribune* was still publishing reports of frozen discoveries. Meanwhile, Minneapolis mercury hovered at twenty-eight degrees below zero.

In the days before weather forecasting, radar, and television, the storm caught the entire state by surprise and had left all but a few journalists speechless.

"One of the worst storms known in years came down upon Minnesota last night," the *Winona Republican* newspaper reported, "like a wolf on the fold."

THE WILD, WILD WEST
OF MINNESOTA

- 1876 -

THREE MEN SAT EATING BREAKFAST in a restaurant in Northfield on September 7, 1876. They had attracted the attention of the towns-folk throughout the week. The little college town noticed the strangers' fine dress; perhaps they came from the Twin Cities. And, their amaz-ing horses drew more than a couple gawkers, just like a fine sports car would a century later.

They weren't from around Northfield, and they probably weren't who they pretended to be. Throughout the week these three men and a couple others with them had tried to pass themselves off as railroad engineers, farmers, and cattle dealers, but the townsfolk didn't buy their stories.

The trio ordered a breakfast of ham and eggs, asking for four eggs each. They ate leisurely and discussed politics lively enough for oth-ers to overhear. They all seemed to be staunch Democrats, unusual for the young state which tilted toward the Republicans. One of the men bet the restaurant owner one hundred dollars that Minnesota would vote Democratic in the upcoming presidential election. The restaurant owner shrugged off the bet and the men left.

Four days earlier, on September 4, 1876, the same three men, along with five others, rode into Mankato, looking to rob the First National Bank. A couple of them had been to the bank a few days earlier when they went inside to change a fifty dollar bill, most likely an excuse to case the surroundings. When they returned to Mankato on September 4, it seemed as if the entire town was on the streets, ready to thwart their attempted robbery. The men fled town quickly, looking for another less vigilant target.

Entirely by coincidence, the Mankato streets were full of people on September 4 because the town had called a meeting of the Board of Trade to promote local business. It was a sort of sidewalk sale. The residents hadn't suspected that a bank robbery was about to occur.

The group of bandits was known as the James-Younger gang, and they are credited with practically inventing bank heists. Eight of them had converged on Mankato and, later, Northfield. They were Frank and Jesse James; Cole, James, and Robert Younger; Clell Miller; William Stiles; and Charlie Pitts.

Before the attempted heist in Northfield, the James-Younger gang met in the woods to put the finishing touches on their plans. They were noticed there, too. One farmer saw the gang of men playing cards in the morning. He later reported that one of the men—in all likelihood, Frank James—had inquired about where to buy "cee-gahs." If the cosmopolitan clothes were not enough, if the stunning horses didn't attract enough attention, the deliberate drawl of a Southern accent surely was noticed.

Shortly after two in the afternoon, three of the men showed up in front of the dry goods store. Not long after, another two entered town from a different direction. The other three placed themselves around town in strategic positions. The townspeople noticed the men and noticed their long duster coats. Underneath the dusters, the robbers were "armed to teeth."

The first three men entered First National Bank. J. S. Allen, a merchant, attempted to enter the bank shortly after the three men. Allen was told to stand back, and he quickly understood what was happening. He ran through the streets, sounding the alarm.

"Get your guns, boys! They are robbing the bank," Allen cried.

Dr. Henry Wheeler heard Allen, grabbed a gun, and positioned himself in the Dampier Hotel, across the street from the bank.

Inside the bank, the robbers wanted the safe opened. They first set their sights on J. L. Heywood, who denied he was the cashier. The gang questioned the other two employees and came back to Heywood, whom they insisted was indeed the cashier.

Meanwhile, outside other members of the gang told townspeople on the street they would kill anyone who didn't get off the street and take cover. A Swedish immigrant who didn't understand English was shot in the head when he didn't obey.

And then the story takes a drastic turn from the stereotypical Wild West bank robbery story: The citizens began to fight back.

The gang members quickly found themselves the targets of the townsfolk, who armed themselves and began to shoot. Allen happened to be a hardware dealer and kept a stock of guns on hand. A. R. Manning, another hardware dealer, joined in the battle. The situation had turned on the James gang. Some town members started pelting the robbers with rocks.

Inside the bank, the situation turned worse. The robbers took some change, mostly nickels, but still had not managed to make it inside the safe. Heywood's lack of cooperation was causing the gang trouble.

"You are the cashier, open the safe quick, or I'll blow your head off," one of the robbers reportedly told Heywood.

The bandits grabbed Heywood, who was the acting cashier. He was simply filling in for the regular cashier who was out of state.

"Open the safe now or you haven't a minute to live," the bandits told him.

Heywood started screaming, "Murder! Murder," and said the vault was on a timer and couldn't be opened. One of the robbers took his .44 caliber revolver and struck Heywood in the head, demanding the safe be opened. Heywood refused.

Pitts took the gun, placed it against Heywood's forehead, and fired, killing Heywood instantly. At that time, Alonzo Bunker, a teller, managed to escape out the back.

The book, *Robber and Hero: The Story of the Northfield Bank Robbery*, records an eerie conversation between Heywood and the president of Carleton College, which is located in Northfield, just a few days prior to the robbery, "Now if robbers should come in here and order you to open this vault, would you do it?"

Heywood answered, "I think not."

Outside, the barrage of bullets got heavier, according to eyewitness accounts. One of the gang members yelled, "The game is up! Better get out boys."

It took only seven minutes for the streets to be lined with dead horses, dead men, countless bullet holes, and shattered windows. The robbers only managed to take $290, a paltry sum for such a big calamity—one that left Clell Miller and William Stiles dead. They were killed by townspeople who fought back.

Almost as soon as the robbers left the city limits, a posse had formed and was following closely behind. The gang was in trouble. Cole Younger had been badly wounded. They went south, heading for the thick woods that ran from Minnesota to Iowa. Although telephones didn't exist yet, news of the bank robbery spread immediately via telegraph. Northfield wired the governor and by night's end, whole trainloads of men were coming from around the state to help search for the robbers. Once news of the bank heist began to trickle into newspaper offices, groups gathered outside the local newspaper offices, waiting for the latest bulletin.

For several weeks, over a thousand men searched for the injured robbers. On September 21, the Younger brothers and Pitts were spotted near Madelia. Bob Younger's elbow had been shattered by a sharp shot from Dr. Wheeler. He had been unable to travel, and his brothers and Pitts refused to leave him. Their cover was blown when two of the robbers confronted a seventeen-year-old Norwegian immigrant named Asle Oscar Sorbel, who suspected them of being the robbers.

Sorbel quickly notified the authorities. Watonwan County sheriff Jim Glispin gathered a group of six men, who approached the hideout. A brief, but fierce skirmish took place. Bob Younger was hit in the chest, James Younger received five different gunshot wounds, Cole Younger had eleven, and Pitts was killed, having been hit five times. The only member of the sheriff's posse to come close to being injured was W. W. Murphy, who was hit by a bullet that ricocheted off a briar-root pipe in his pocket and lodged in his pistol belt.

The posse took the Younger brothers into custody and later a grand jury returned four indictments against them. They pleaded guilty to the charges, avoiding the death penalty. In Minnesota, a death penalty could only be enforced if a jury ordered it. Since the Youngers did not go to trial, they avoided a date with the gallows.

The Youngers, who had wreaked such havoc on the civilized world, became respectable while in prison. They became so well-known that people across the state held annual campaigns for a governor's pardon. Even Heywood's widow and daughter signed petitions to have the Youngers, who plead guilty to being accessories to Heywood's murder, pardoned.

Bob Younger died in prison from tuberculosis in 1889, but not before he became known for his diligent study of medicine. He also started the oldest prison newspaper in America, the *Prison Mirror.*

In 1901 legislators impatient with the governor's reluctance to set the Youngers free, proposed a law that those serving life sentences could be freed after serving thirty-five years. By then the Youngers had served about twenty-five years and received eleven years credit for good behavior. Despite protests from the Northfield legislative delegation, the law passed, and the Youngers were released with conditions.

On July 14, 1901, the Youngers were released, but they could not leave the state, nor could they appear as a feature in a show or exhibition. Cole and James Younger both worked as salesmen in the Twin Cities for a time. They became students of the ever-changing world around them and were anxious to see the inventions of the twentieth century.

"[The telephone] tickled me to death, because I had heard them talking at one end of the line and it was all I could do to keep my face straight at the spectacle of a fellow jabbering into a dumbbell," Cole said, before his release from prison in a newspaper interview.

James Younger committed suicide in the Reardon Hotel in St. Paul. He was upset by the Board of Pardon's decision not to let him marry.

On February 4, 1903, the board granted Cole Younger a pardon with a catch. He would remain a free man as long as he never returned to the state. He went to his home in Missouri, where he found his old partner, Frank James, in a Wild West show which toured the United States. He made several unsuccessful attempts to return to Minnesota, but the board always rejected those requests.

Cole Younger died in 1916, after fifteen years of freedom. Frank James had died the year before, at one time standing accused of robbery and murder, but never doing time for those crimes.

THE END OF AN ERA:
MINNESOTA IS SETTLED

- 1878 -

One eager journalist, covered in dust and dirt, rode ahead of a lumbering oxcart train as it wound its way over the Red River Valley. With his head down and his hand feverishly scribbling on his note pad, he chronicled this scene for the (Winnipeg) *Nor'-wester* in 1860:

> *Looking back at the line extending far over the plain,*
> *the spare cattle following, and horses galloping about*
> *with very Cossack-looking prickers after them, and the*
> *train winding its way, like a great snake, coming along*
> *very slowly, and then, when it got near enough, they*
> *hear the carts which, at a distance sounded not un-*
> *musically, and then the lowing of the cattle and the*
> *songs and voices of the men.*

The now forgotten writer told it like he saw it, not like it really was.

Samuel H. Scudder, writing about the carts years later, said, "Each cart waggles its own individual waggle, graceless and shaky."

By all accounts, the unique oxcart developed by métis fur traders was both a mechanical marvel and a painfully primitive invention. Without it western Minnesota might not have been settled at all. For three-quarters of a century, the carts linked the remote land to the rest of the world. They were an invention brought about by sheer necessity and had the character of rugged frontier transportation, not unlike a covered wagon or a paddle-wheel steamer.

The wooden wheels squeaked incessantly as they rubbed against the axles without lubrication. "Each wheel was said to have its own shriek," wrote historian John Caron.

An 1878 article in *Harper's Magazine* joked, "The axles are never greased and they furnish an incessant answer to the old conundrum: 'What makes more noise than a pig in a poke?'"

The wheels went unlubricated because dust and dirt from the plains would mix with the grease causing it to work like sandpaper, damaging the cart. Despite their noise, these carts were nothing if not Minnesotan—sturdy, practical, and without luxury.

The cart's construction was crude, but thoroughly functional. The two wheels were over five feet in diameter, with wooden spokes angling out from the hubs to the rims. The huge wheels gave the cart clearance and stability. Everything about the cart was built for the harsh plains which were like marshes in the wet spring, ovens in the summer, and prone to blizzards in the fall. The rims could be up to three inches thick in order to keep the wheels from bogging down into the soft sod. The rims consisted of six curved sections called "felloes."

The spokes, usually a dozen, were mortised into the hub and completely through the felloes. Maybe the most amazing part of the simple cart was the axle that supported the entire load which could range from eight hundred to a thousand pounds, according to various sources. These wooden, ungreased axles would, and often did, fail, sometimes several times per trip. According to the Clay County (Minnesota) Historical Society, a cart traveling from St. Paul to Winnipeg would go through four or five axles per trip. Most travelers had a solution to the common problem as they would simply lash spare axles to the rear of the cart. The axles were cone-shaped at each end to fit into the hub of the wheel where a lynch pin would lock the axle into the wheel. If any part of the cart broke, it could be replaced en route because the entire contraption, from wheel to box, was made of wood, making any nearby tree the pioneer equivalent of an auto parts store.

Cart travel remained brisk for over a half century extending even after the Civil War. Oxcart trains set out from urban centers like St. Paul and St. Cloud and headed for Red River Valley up through Canada. Any and all supplies had to be transported by these crude carts, which could stretch for as far as the eye could see.

The carts ushered in a golden era of traders, charlatans, characters, pioneers, and politicians. In other words, the carts ushered in a rough-hewn pioneer civilization.

For example, there was "Jolly Joe" Rolette, "King of the Border." Rolette "never neglected to cultivate the legend of himself," as historians Rhonda and Carolyn Gilman write. Rolette sang Ojibwa songs, gave impromptu dances, and wore American Indian clothes, even though his ancestry was French and British, the nearest Indian blood being a great-great-great grandmother, who was a member of the Ottawa tribe. Like other traders—notably Henry H. Sibley—Rolette would become a force in Minnesota politics. As a legislator, he stood opposed to a law that required the bonding of all the territory's free blacks. And, Jolly Joe would be the person responsible for ensuring that St. Paul remained the state's capitol.

Jolly Joe's opposite also traveled the Red River trails. A dark, six-foot-tall quiet trail master, Pierre Bottineau had a face with angles as sharp as a knife and a piercing stare to match. He earned a reputation as a master guide who traveled the region for nearly three decades. He accompanied railroad surveyors, took emigrants from Minnesota to Montana during the early days of that state's gold rush, and accompanied Sibley's military forces while the Dakota warred with settlers. Bottineau left no memoirs of his life. However, trader Martin McLeod remembered once hiring the young Bottineau as a travel guide.

Bottineau's party of four was in North Dakota in springtime when a blizzard hit hard. The twenty-year-old Bottineau rescued McLeod, who had been separated from the party. Bottineau saved his life by "using blows from his rifle butt to keep the exhausted and freezing McLeod on his feet." McLeod would fare better than his other two inexperienced companions. Despite efforts to save another man, only McLeod and Bottineau survived.

And, men like James McKay, who resembled a grizzly bear more than a human, prowled the Red River trails. McKay, born of American Indian, Scottish, and French descent, would later speak several Indian languages and become a fixture in Manitoba politics. A contemporary described him as "a man of enormous bulk and catlike grace." One traveler recalled that in a St. Paul hotel, "Yankees in shiny boots creaked and stomped about like so many busy steam engines," while McKay, probably wearing moccasins, paced the floor without making a sound.

The carts were as unique as the characters who drove them, and just as suddenly as the oxcarts had appeared, they disappeared. With them disappeared one of the last vestiges of a rugged new world. Even though the shriek of the cart has vanished, memory of them has not been completely erased.

Today, trail ruts can still be seen running along fields and across the now-settled plains.

BURNING DOWN THE HOUSE:
THE STATE CAPITOL BURNS

- 1881 -

WITH JUST A DAY LEFT TO GO, many Minnesota state senators milled around the back of the Senate chambers, hobnobbing and speculating about the bonding bill which was being debated. It was always this way—a few late nights and a flurry of bills to get finished before the legislature adjourned. Reporters and spectators had gathered, despite the late hour, to watch the final, eventful moments of the session. All was quiet at the statehouse.

Shortly after 9:00 p.m., March 1, 1881, someone alerted the Senate that the capitol was on fire. When the senators realized what was happening, they turned to see the only stairway out of the building engulfed in smoke and flames. At that moment, someone burst into the chambers and rather unceremoniously yelled, "Fire!"

Three hundred people snapped out of shock and into panic as most rushed to the staircase, only to be beaten back by the heat.

Lieutenant Governor Charles A. Gilman, serving as the president of the Senate, remained calm and after a few men were knocked back by the flames suggested, "Shut the doors. Shut the doors. Don't make a draft. Act like men."

As the panicked crowd obeyed, one senator moved to adjourn, and it was quickly voted on unanimously.

Gilman probably saved his fellow senators and spectators. Closing the doors did seem to prevent a draft and the flames subsided long enough for the crowd to make a dash for the stairway. A couple of people in the Senate chamber ran out onto the veranda where they started yelling for ropes and ladders. A few discussed the possibility of making a ladder out of the carpet.

While the senators rushed out of the building and onto the veranda, the clerks, with little agitation, picked up their records and some bills, wrapped them up, and quickly exited the building with a lot less excitement, saving some of the work that had been done during the session.

In the other chamber, the House continued its business. The representatives first sensed something was wrong when cinders fell from the ceiling. At about the same moment, a black cloud of thick smoke rushed into the chamber. The lower chamber didn't fare much better than the Senate. The only escape for two hundred people was through a long, narrow, smoky hallway which extended for nearly one hundred feet. A few people jumped out windows, opting for ice rather than fire. One Washington County representative took the plunge out a window to land in a snow bank "with a few unimportant bruises."

Like their counterparts in the Senate, the clerks rather unceremoniously gathered the records and carried them to safety.

Some newspaper accounts said the time from first alarm to a raging inferno was less than three minutes, others said it was more like seven. The blaze lit the city in a glow as the capitol was engulfed by flames. Firefighters could do little more than watch the structure burn. By morning, the building was a charred heap, as were most of the historic records inside.

Estimates place the number of people inside the capitol at the time of the fire at around five hundred, and all managed to escape without major injury. The value of the building itself was estimated at eighty thousand dollars. The building had been the first state capitol, although two other structures had served as home to the territorial legislature prior to its construction in 1853.

Curiously, the state had never insured the property. All that was found in a safe was a single ten thousand dollar insurance policy from London, Liverpool, and Globe Company for the library.

Much of the state library, which was also located in the capitol building and had held nearly thirteen thousand books, was lost. Some old, rare books were salvaged, but fewer than a thousand were saved. The library's loss was pegged at seventy-five thousand dollars. Charles Chappel, a one-armed janitor worked tirelessly during the fire to save some of the treasures of the state library, including a set of old English reports, said to be among the most valuable books in the library at the time. He kept working until a burning beam struck him on the head. Taking that as a sign that he should get out, he grabbed a few more books and rushed to exit the inferno, saving a little bit more of the state's library.

The state's supreme court records were hauled out as men fought the fire. The state historical society records were also saved and moved across the street to a Universalist church. The Academy of Science's room in the basement of the capitol, however, couldn't be reached in time. In the panic and chaos of the fire, the keys to the room were lost, and a newspaper lamented that "a very valuable collection of natural curiosities is therefore lost."

With nothing more serious than a few bruises, the governor

escaped the capitol that night and the *Minneapolis Tribune* reported that "Governor Pillsbury would be justified in making modest provision for the replacement of the several dozen pairs of coat-tails which we regret to learn were burned off before the people's representatives could secure exit from the late lamented temple of law."

Others didn't find the same levity in the situation.

"I had made up my mind that my time had come and said as much to Mr. Dunn. I've been in tight places before, but never in one that seemed to threaten so sure a fate as that for a few minutes," said Colonel L. L. Baxter, who witnessed the fire.

THE TOWN SO NICE IT BURNED TWICE

- 1893 AND 1900 -

FIRES FLARED UP AND RAGED around the city of Virginia throughout the spring and early summer of 1893. For three weeks, blazes consumed the thick forests near the town, and the residents had grown used to the smell of smoke. No fire had ever really threatened the city itself.

But on June 18, 1893, fire broke out in the town of Virginia, but it seemed to excite little reaction in the citizens. At first they were hesitant to call for help, because Virginia was a mining community and help might just bring folks who would try to hone in on the prospectors who were already there. The fire erupted at around one o'clock in the afternoon, and by two o'clock, one person wired, "For God's sake send us aid."

By six o'clock that night, the little town of Virginia, including its forty-two saloons, five hotels, and five hundred houses, was destroyed.

Amazingly, not a single life was lost, although many who tried to fight the inferno were burned. Residents reported that as soon as the fire started to spread, people started looking for loved ones and the entire town made an instinctive rush for the train station, where trains began transporting citizens to Duluth.

As for the cause of the fire, "the people of the village, too, were somewhat at fault," said John Schultz, a lumber dealer, in a newspaper article. "The town has been so long in danger that they ceased to appreciate the gravity of the situation. They were careless when the end came and not ready to meet the terrible emergency."

Another man, Hermann Nichols, quipped, "I went into Virginia rich and I come out poor."

Shortly before two o'clock, residents noticed that a wind, what one described as a "perfect gale" fanned the flames, ensuring that few buildings in Virginia would be saved.

Accounts of the fire also include mention of the "hero of Virginia." Bradley Taylor learned that a young woman was still in a building that was on fire. While others stood by watching, Taylor, who was reportedly ill with a fever, grabbed a ladder and went to get the young woman. When he returned to the ladder with the woman, the flames had literally burned away the ladder. By this time, the crowd had snapped out of its daze and went to get another ladder for Taylor. Placing a blanket around the woman, he carried her down the ladder to safety. For his efforts, Taylor was badly burned and his eyes soon swelled shut. He was evacuated by train to Duluth, where he arrived to a hero's welcome.

When evacuated Virginia citizens arrived in Duluth, they found that the city was ready to help them. Duluth had already started a collection for the evacuees and restaurants immediately offered sandwiches, cakes, pies, and coffee. "The tables were what first attracted the hungry crowd. They could not stop to talk. Sandwiches, coffee— anything to eat—disappeared faster than it could be supplied," according to several newspaper accounts.

Almost immediately, the rugged spirit of Virginia's citizens began to surface. As soon as the damage had been assessed, city leaders began making plans to rebuild the town. Still, many wondered how the gritty little city would be rebuilt.

"Trouble is feared at Virginia from the character of the population," a reporter in Duluth wrote. "As in all new mining towns, the men are many of them hard characters, and lawlessness is anticipated. Police Sergeant Smallett and a squad of officers have been sworn in as deputy sheriffs and will be sent to Virginia tomorrow morning."

The town rebuilt only to have history repeat itself seven years later.

On June 7, 1900, two fires broke out, destroying most of the business district, but leaving one saloon, a grocery store, and a meat market.

Two thousand residents—about the same as seven years before—were again left homeless.

The first fire began at about 11:20 a.m. at the lumber mill, owned by the Moon and Kerr Lumber Company. A blower had caught fire and the sparks spread rapidly. About forty-five minutes later, Ole Halversen's butcher shop in the middle of town caught fire. The shop sent out a call for the volunteer fire department, but the volunteers were busy at the lumber mill, saving two hundred thousand dollars worth of lumber while the rest of the town burned.

It took four hours for the city of Virginia to burn. Ten city blocks were destroyed by late afternoon. Fire had destroyed the town again, and amazingly, just like seven years before, not a single person perished.

By the end of the day, Duluth had already stepped up and was again taking collections and sending relief.

In the same way that fire rejuvenates a forest, Virginia's fire sparked new growth in the city. In the month that the fire occurred, June 1900, the population hovered around twenty-seven hundred people. Five years later, it surpassed six thousand. Thirteen schools and many sawmills, mines, and railroads sprang up around the twice-burned city between 1904 and 1913.

TWO TRAINS TO SALVATION:
THE HINCKLEY FIRE

- 1894 -

THE SUMMER OF 1894 HAD BEEN CRUEL to Minnesota. The sun beat down on Hinckley and its residents. Clad in wool and linen, they could find nowhere to hide from the oppressive summer sun. The land was parched, and there was no shelter, no shade—stumps sat where trees once stood. This was timber country, and scenery had given way to progress. Residents looked out their windows and saw a sea of stumps. They could also see hills of logs waiting to be turned into lumber for a nation just waiting to be built.

Hinckley and the Brennan Lumber Company boasted their new firefighting equipment and the state-of-the-art fire suppression at the lumber mill. The new equipment was required for protection from fire in the middle of Minnesota timber country.

However, no Hinckley homeowner could get insurance on a wood-framed house in the middle of a forest or in a town that had thousands of board feet of lumber stacked within city limits. Firefighters might be able to ward off a small blaze, but the questions lingered about whether the small community was able and prepared to fight anything larger than a mattress fire.

Soon that question would be answered as a fire began to creep toward Hinckley late one afternoon. Fire wasn't uncommon in timber country, and Hinckley residents had faith in the new firefighting equipment. A half an hour later, a few dozen buildings on the outskirts of town had begun to burn. About that time, fifteen miles outside of Hinckley, train engineer Edward Barry noticed thick, dark smoke.

Smoke wasn't anything new in this timberland, especially in a drought year. The engineer had the crew light the head lamp and cab lights so that Engine Number 24 could make it safely into Hinckley. Ten minutes later, at 2:40 p.m., the Great Northern train pulled into Hinckley. Barry couldn't believe what he saw. The entire town seemed to be on fire.

Barry pulled the train onto the sidetracks. He could see that other boxcars were on fire, and even the rails were beginning to warp and burn. Barry was left with two grim options—leave and be set on a certain collision course with another incoming passenger train on the single track, or wait until it arrived at the scheduled time some forty minutes later. Barry and his crew opted to wait for what must have seemed like an eternity while the town burned around them.

Meanwhile, engineer William Best pulled out of Duluth at 1:00 p.m., headed for Hinckley. Thirty miles south of Superior, Wisconsin, Best lit the headlamp and cab lights because darkness from the smoke engulfed the train. He thought for certain the train would be wired a message if there was a horrible fire. A message never came, so Best continued in Engine Number 4 toward Hinckley. As the train drew near the city, the sky suddenly became brighter, but the light didn't come from the sky; it seemed to be shooting upward through the columns of smoke at the south end of the lumber town.

Amazingly, the train arrived on time, but within moments, the crew realized it would be impossible to push south toward the rest of the stops because of a wall of flames at the end of town, which blocked its path.

The two engineers, Best and Barry, agreed to hook their trains together and retreat northward, away from the inferno. The flames were so intense and the tracks were so hot that crews couldn't get the turntable to turn the engines around, so the two conjoined trains would have to travel out of the city in reverse. The combined crews hooked three empty boxcars to the train to carry Hinckley residents, who were coming to the trains from all directions. The crews helped the residents, most of whom had only the clothes on their back, onto the trains.

With the flames burning hot and ashes falling from the sky, the engineers were caught in heated disagreement. Barry wanted to leave, but Best applied the brakes as more people climbed aboard. Despite Barry's attempts to power the trains forward, Best continued holding the brakes. Finally, he let off the brakes when flames started to ignite the tracks below the cars. About 350 Hinckley residents had made their way onboard.

Best couldn't forget the residents who came running after the train as it pulled out of the station at 4:00 p.m., nor could passengers forget the man who tried to catch up to the train by riding his horse alongside the passenger cars. Passengers reached out to grab him, but at that moment, his horse pulled away, straight into the firestorm. Outside town near the Grindstone River, the train stopped briefly to pick up a few residents. Days later, recovery crews would find many bodies of people who had journeyed to the Grindstone, following the route the others had, only to be burned to death.

The two engineers feared neither the tracks nor the bridges along the way would hold. As they traveled along, they couldn't help but notice the railroad ties on fire. Windows shattered because of the heat, and cinders caught passengers' clothing and railroad cars on fire. Inside people cried, prayed, and some boasted they had guns and would use them to commit suicide rather than be burned to death. One man begged for morphine.

Farther down the tracks near Sandstone a large bridge—the Kettle River high bridge—rose over 150 feet in the air. The crews could see that the bridge was on fire. They didn't know if the bridge could hold the weight of the trains. They started over the bridge at the recommended speed of four miles per hour, reaching the other side safely just minutes before the structure collapsed.

Nearly six miles later at the Patridge station, the train took on more coal and water. The crew went throughout the train with water buckets, trying to help the burned passengers.

Both Best and Barry remained at the helm of their engines until they reached their homes in Superior. Barry's eyes were so badly burned that he could barely see. Because of their bravery, 475 people arrived safely in Duluth before 10:00 p.m.

All that remained of the lumber towns was cinders and bodies. Crews worked to fix railroads, bridges, and eventually the structures. Several towns, including Hinckley, would slowly be rebuilt from the ashes. On September 10, Hinckley city council members gathered at the ashes of city hall. All of them had survived the fire, but Mayor Lee Webster had lost his wife and parents. Their first order of business was allocating money for iron cells that would hold criminals who interfered with the returning residents.

Within a month, white tents lined the city streets, and rebuilding had begun. Hotels, restaurants, saloons, and permanent houses followed not long after. Two months later, more than a hundred homes started to reappear, making Hinckley like the mythical phoenix that rose from the ashes, beautiful and new.

THE DISCOVERY OF THE
KENSINGTON RUNE STONE

- 1898 -

As OLOF OHMAN WORKED CLEARING a poplar tree from a field outside of Kensington on an early November day, he stumbled across a
heavy mica stone with strange marks on it. The stone was three feet
long, a half-foot thick, over a foot wide, and weighed over two hundred pounds. Ohman thought it might be some artifact from an
American Indian tribe.

Ohman and his neighbor, Nils Flaten, thought the stone might
indicate buried Indian treasure. After some futile digging, they gave
up on their hunt for buried treasure and took the stone to town,
where it gained attention in the window of a bank. Ohman, a native
of Sweden, and Flaten, born in Norway, didn't realize the stone had
more to do with their own heritage than buried Indian treasure.

Eventually, the stone made its way to the University of Minnesota for research. The strange carvings were identified as runes,
letter-symbols of old Swedish, the language of Norse explorers.

Before experts could finish studying it, newspapers and magazines were hailing the Kensington Rune Stone as one of the most important archaeological finds in North America.

There was one problem. Academics and experts simply couldn't agree whether the stone was proof positive of European explorers, who had come deep into the interior of America over a hundred years before Christopher Columbus, or whether it was an elaborate hoax. The stone sparked an immediate sensation in the rather arcane world of rune study, with a majority of professors denouncing it.

While the message on the stone is undoubtedly in rune and while it has a believable message, too many other problems cast a pall on its authenticity.

Most scholars believe the runes can be roughly translated:

> *8 Swedes and 22 Norwegians on an exploration journey from Vinland westward. We had our camp by 2 rocky islets one day's journey north of this stone. We were out fishing one day. When we came home we found 10 men red with blood and dead. AVM [Ave Maria] save us from evil. We have 10 men by the sea to look after our ships, 14 days' journey from this island. Year 1362.*

The stone was declared a hoax by academics and was sent back to Ohman in Douglas County. Seemingly unfazed by the criticism, Ohman took the slab back and used it as a stepping-stone in his granary.

But it wouldn't be there for long, because Hjalmar Holand, a professor at the University of Wisconsin, took up the stone's cause again. He overcame many objections from his contemporary critics, including explaining how the Viking exploring party could have made it to Minnesota in just two weeks. Holand wrote books and articles about the stone, which also brought forth an impressive flurry of critics.

The academic debate, however, didn't seem to deter interest in the stone, as it was displayed at the Minnesota State Fair, the Chamber of Commerce in Alexandria, and even at the Smithsonian, which originally gave the stone its imprimatur of authenticity, only to withdraw it a couple of decades later.

Today, the stone is still controversial, and every couple of years a new scientist, linguist, or historian unearths a fact which he or she believes settles the argument once and for all, only to have it bring the same cacophony from believers and skeptics.

BLACK MASS

- 1915 -

AROUND 8:45 ON A FRIDAY MORNING in August 1915, Bishop Patrick Heffron was celebrating Mass in a small chapel at Saint Mary's College, an institution that he helped found. As he said Mass, he faced the altar. He didn't hear the man walk into the chapel behind him, nor did he sense the man's approach.

The man, wearing a double-breasted Prince Albert suit and gripping a revolver, took aim at the Winona bishop and pulled the trigger. The first shot hit Heffron in the left thigh. Heffron wheeled around quickly, just in time to see the man pull the trigger again. This time a bullet struck him in the right side of his chest. The assailant fired once more, missing the bishop.

Then the man fled the chapel. The shots had drawn the attention of the other priests and college staff, who came running toward the chapel. As several priests rushed in to help the injured bishop, others called the authorities and summoned medical help.

After fleeing the chapel, the assailant, Father Lawrence M. Lesches, headed to room 135, where he'd been staying for several days, and locked himself in the room.

Winona County sheriff W. E. Parr received the call about the shooting at the college chapel at 8:50 a.m. He sent his son, Willis, out to get the car, and then he called George Huck, the Winona chief of police. The trio set out for the college, six miles outside city limits.

By the time they arrived, the priests had identified Father Lesches as the assailant. He had been seen leaving the chapel right after the shooting. The priests knew that he was in his room and had heard him flush the toilet repeatedly. The sheriff told his son to stand guard outside the building in case Lesches tried to escape through the window. Meanwhile, Huck and Parr approached Lesches's room, revolvers drawn. Parr stood to the right side of the door, and Huck stood to the left. Parr rapped on the door three or four times. Finally, they heard the lock rumble and the door began to open. Wasting no time, they pushed their way inside. Lesches wasn't armed and was taken into custody immediately.

By 9:25 a.m., just forty-five minutes after the shooting, Lesches was in a jail cell. He had traded his Prince Albert suit for a khaki jail uniform. He had $5.12, a pocketbook, a small knife, and a few other items on his person when he was booked.

Meanwhile, two doctors were summoned from Winona to look after Bishop Heffron. Dr. William J. Mayo, one of the two famous Mayo brothers, was also called, and he made the two-hour trip from Rochester to Winona. The second bullet caused the most serious injury because it struck the right side of Heffron's chest, right above the fourth rib. The bullet then penetrated the lung and lodged somewhere behind it.

The first bullet, which struck the thigh, was a flesh wound and was not serious. All three shots were fired close enough to the bishop to leave gunpowder marks on the bishop's vestments. At first the doctors were concerned that Heffron's chest wound would become infected. They waited three days, then declared that Heffron would likely make a full recovery but that he would have the bullet lodged behind his lung for the rest of his life.

At first it looked as if Lesches's motive to shoot Heffron was simply revenge, spurred by a confrontation a day earlier. Later it was determined that Lesches was mentally disturbed. Lesches told reporters just hours after the shooting that he'd met with Heffron the day before and the bishop had refused to give him an assignment. In a written statement Lesches recounted that meeting, "I cannot go further. I have no money, clothes nor friends." Lesches said the bishop had replied, "I don't care. I'll see that you get out of here."

On the evening before the shooting, some priests were in a common room playing cards. They noticed Lesches was pacing and acting agitated. Lesches asked Reverend Bernard Kramer what he would do if the bishop refused to assign him to a parish. Kramer jokingly replied, "I'd shoot the son of a bitch." Kramer later worried that his comment gave Lesches the idea to shoot the bishop.

Heffron had had doubts about Lesches for decades, since the time when Heffron was a seminary headmaster and Lesches was a struggling priest in training. By the time Heffron became a bishop, Lesches's mental problems had grown severe, causing Heffron to move him frequently. Each time Heffron reassigned Lesches, the priest begged for more money, threatened harm, and even hired an attorney to sue the bishop for neglect.

Heffron recovered from his wounds and a trial began with the bishop as star witness. Meanwhile, the diocese, which had been somewhat reluctant to give funds to Lesches prior to the shooting, agreed to fund his defense.

On the stand, Heffron admitted that he had had serious misgivings about Lesches's ability to lead a congregation. The jury found

Lesches not guilty by reason of insanity and sent him to St. Peter's Hospital for the Insane, which was, ironically, the same place that Lesches's erstwhile attorney had suggested the church send him almost a year before the shooting.

In his own way, Lesches tried to apologize for his actions. After a year at St. Peter, he wrote a letter to Heffron that was both remorseful and accusatory. "It is with pain and great sorrow that I come to apologize for the unfortunate deed which must have plunged your heart in deep affliction . . . you refused me food, clothes, shelter and care in sickness and helplessness; you said you did not care for my life or soul, to find work with some farmer and you knew I was crazy. What was then left to me but death in the gutter or insanity?"

In 1919 the state of Minnesota briefly considered sending Lesches back to France, where he had a brother who would care for him. That plan fizzled, maybe due in part to objections from Heffron, who worried about what Lesches might do if he was released from St. Peter's.

Lesches did manage to outlive his antagonist by over a decade. Lesches died at St. Peter's Hospital in January 1943. Ten priests attended the funeral. Inside the walls of the hospital, Lesches had found some measure of peace, taking communion, taking an interest in the lives of young people in the church, and earning the respect of those who worked there.

WALKING ON WATER:
WATERSKIING IS INVENTED

- 1922 -

RALPH SAMUELSON WAS A FIXTURE ON LAKE PEPIN. While his father worked hard at the family grocery store just to make ends meet, the lithe, suntanned Ralph splashed in the cool waters of Lake Pepin, diving for freshwater oysters, which the local button factory bought for their pearly enamel. He knew every rock and every inch of sand on the beach. And in his mind he could trace every peak of the bluff lands surrounding Lake City.

Once, while diving in the lake, he found a rare freshwater pearl, but that wasn't his most important discovery at Lake Pepin. During the winters, he'd slide down the riverbanks and bluffs on barrel staves. And, that's where the idea first struck him: If he could slide down bluffs, then couldn't he do the same on water? Couldn't you glide on water?

"Everybody thought I was completely nuts," Ralph said.

And, for awhile, everybody appeared to be right. In late June Ralph tried to ski on water.

It wasn't the first time he tried to grandstand. Ralph's brother, Ben, would hitch a flat board up to his boat, and Ralph would perform tricks and stunts to the delight of a crowd in search of cheap summer entertainment.

Naturally, a crowd turned out to see Ralph attempt to ski on water, something that hadn't been done before on any of Minnesota's other 9,999 lakes or on any other lake in the country.

Ralph first used snow skis, reasoning that if they were good enough for ice and snow, then they'd be good enough for water. Ben fired up the six-cylinder boat, and Ralph proceeded to take in a lot of water and sink.

Undeterred, Ralph searched for his next pair of water skis, this time finding barrel staves. Again, Ben revved the engine, and again Ralph sank. By this time, Ralph had put on quite a show for the crowd on the shore. Many shook their heads, muttering that he was just plain nuts.

Still unfazed, Ralph went to a lumberyard, where he bought two boards of lumber for a dollar each. The boards were bulky and wide, measuring nine inches wide and eight feet long. Before taking them out for a spin on the lake, Samuelson used steam to soften the corners of the boards, and with a vise grip, gradually curve the ends. With his homemade skis, he gave skiing on water another try. The skis promptly proceeded to snap.

Ralph believed he was getting closer to success. He could almost taste it, along with a lot of Mississippi River water. He went back to the lumberyard and bought two more planks, this time ten inches longer. Ralph had an ironworker make rings which reinforced the ends and added some rubber and foot straps to help him stay on the unwieldy boards. Ralph also had a local blacksmith forge a ring that he wrapped in rubber and attached to a rope that would be harnessed to the boat. This would make the rope stronger and easier to grip.

He first attempted to start skiing off a ramp. This proved to be a human rock-skipping demonstration. Ralph jumped off the boards, skidded along the surface of the water, and plunged into the lake.

The idea—to skim along the surface of the water—seemed so simple, yet it was proving remarkably hard. After days of trying, Ralph tried skiing on water with a new twist. Instead of starting off on a ramp or on the boards, he'd start with the tips of the boards just barely out of the water. And, after days of bitter, water-logged failure, that slight change was enough to change everything.

The brothers began about a hundred yards offshore in water about ten-feet deep. Ralph skied gracefully across the lake as Ben revved the engine. On July 2, 1922, at around 4:20 in the afternoon, Ralph Samuelson became the first person to successfully ski on water. Now, instead of thinking he was "completely nuts," onlookers had new questions. They now asked Ralph how he did it and if he could do it again.

Ralph obliged the crowd becoming a local celebrity and, for a time, even a national one.

"All I felt at the time was that at long last I had proved to my family, to myself and a lot of doubting Thomases who had been laughing at me for months, calling me a silly fool, and worse, labeling me "Nutty Sammy," that sort of thing, that I could do what I had set out to do, and that was ski on water, as I had skied on snow most of my life," Ralph later recounted in his biography, *A Daredevil and Two Boards*.

Ralph had fulfilled his vision, but there were still people who doubted him. Before the 1960s, the Water Ski Association had always given credit to the French for inventing waterskiing. When a Twin Cities journalist, who was vacationing in Lake City in 1963, spotted the old wooden skis in the Lake City bathhouse, she put out a call for Ralph Samuelson in the *St. Paul Pioneer Press* and found him promptly in Mazeppa. He had never sought credit for his invention.

The Water Ski Association launched a three-year investigation into the claim that Ralph Samuelson had invented waterskiing. Few documentary sources recorded the event, but an entire town had unwittingly been eyewitnesses to history. The investigation revealed that the French had merely gotten the idea from one of Ralph Samuelson's shows. The first patented water skis were not produced until several years after Ralph skied on Lake Pepin.

THE BIRTH OF BETTY CROCKER

- 1924 -

SOME CHILDREN START LIFE IN A TEST TUBE, but that's nothing compared to Betty Crocker.

She was conceived in a test kitchen and brought to life—born, if you will—in the corporate headquarters of Washburn Crosby Company in Minneapolis.

The story of Betty Crocker began when Samuel Gale, the advertising manager for Washburn Crosby, makers of Gold Medal Flour, received letters from anxious housewives needing advice about cooking. Gale would take these questions to the all-female Gold Medal Home Service staff, who would answer the questions for him so he could respond to the inquiries. Gale, however, never felt comfortable signing his name to the letters. After all, who would trust a man to know his way around the kitchen?

Between the all-male advertising staff and the all-female home service staff, Gale believed the company needed one more staff member who could tackle the customer inquiries. Of course, Gale's bosses weren't thrilled about hiring another staff member. And that's when Gale had an idea, one which would become one of the best-guarded secrets in American marketing.

The perfect applicant was available, and the cost was nearly nothing. In fact, the employee would be an invention.

Gale simply took the last name of Crocker from William Crocker, a former director of the company. To that he added "Betty," a popular, simple, and, above all, trustworthy name. The idea was to have Betty Crocker—a fictitious person—become Washburn's public identity. Gale knew Betty would need a signature for letters. He held a contest for female employees for a unique Betty Crocker signature. The first Betty Crocker signature was provided by a secretary, who produced a plain and tidy cursive signature.

From that point on every letter and brochure would contain the Betty Crocker signature. The newly minted "spokeswoman" worked a little too well, because inquiries increased. The job of signing Betty Crocker's John Hancock soon became too much for one person. The Washburn Crosby Company had to train other staff members on the proper way to sign "Betty Crocker" just to keep up with the demand. This personal touch, even down to the trademark "cordially yours" at the end of the letters, was enough to encourage hundreds of letters from women—and men—to ask about more than just cooking.

And soon Betty became more than just a knowledgeable cook with a neat signature. Her face graced all the company's advertising, and she became a maven of recipes a radio celebrity, and an advocate for modern kitchens.

Soon Betty Crocker was selling neat little wooden boxes full of recipes for the bargain price of 70 cents. For 30 cents more, she'd send twice as many recipes. She also struck gold when she updated the Washburn Crosby Company's popular cookbooks, which had last been published at the turn of the twentieth century. The home service staff updated the cookbooks and Betty signed her name to it. It was an instant success.

The company was inundated with so many requests for cookbooks and recipes, that demand frequently outpaced supply.

An entire field service staff helped answer questions regarding baking failures. They also updated recipes and developed easier methods of baking. For example, the Home Service Department led a crusade against irregular pan sizes because they contributed to so many culinary disasters.

Betty Crocker also became a tireless advocate for modernizing the kitchen. The Washburn Crosby Company test kitchen was outfitted with a refrigerator, followed by an electric range. The sticker price of many of these appliances was more than an automobile and therefore out of reach for many consumers, but Crocker championed these appliances.

Crocker's transformation from celebrity to icon didn't happen in the kitchen, but in the living room. While things like an electric mixers and refrigerators made her popular, it was radio that brought her fame.

In 1924 the Washburn Crosby Company decided that radio might just get Betty Crocker into even more homes. Even the most widely read magazines of the era couldn't come near the size of a radio audience, even on a local Twin Cities scale.

The Washburn Crosby Company bought a struggling radio station—WLAG—for no other purpose than to use it as a vehicle for Betty Crocker. The company renamed the station WCCO, after the Washburn Crosby Company, and on October 2, 1924, Betty Crocker took to the airwaves with "womanly talk."

The first voice of Betty Crocker was Blanche Ingersoll, and speaking as Betty, she announced that every morning, at about 10:45, she'd be stopping by the radio to talk about cooking and homemaking.

Soon housewives and other curious listeners learned, "if you load a man's stomach with soggy boiled cabbage, greasy fried potatoes can you wonder that he wants to start a fight, or go out and commit a crime? We should be grateful that he does nothing worse than display a lot of temper."

Betty Crocker on the radio was an immediate success, and the company soon recognized that she could have national appeal. The Washburn Company also understood that it would be a real marketing coup if families, separated by hundreds of miles, could all hear the same radio program—it could create a legion of loyal customers throughout the country. The only obstacle was getting the same programs to scores of radio stations in one day.

The Washburn Crosby Company solved the problem by expanding the broadcast to a dozen stations, stretching from coast to coast. A different woman read an identical script at each station. Crocker's real identity—or lack thereof—remained a closely guarded secret. The script would come from the sinister-sounding "mother kitchen" in Minneapolis.

While the company didn't admit that Betty Crocker was a highly successful marketing gimmick, it didn't necessarily lie about it either. In fact, the Betty Crocker name was so well-known and trusted that few questioned her identity. The company simply stated that she was from Minneapolis and had a staff of field service employees who assisted her. Even when a 1945 *Fortune* magazine article outed Betty Crocker, few seemed to take notice and even fewer cared.

Betty Crocker had answered too many letters and been heard on the radio too many times. Some even cried when they toured the Minneapolis facility, hoping to catch a glimpse of Betty, and didn't see her. Even when the company revealed the truth about Betty Crocker, some refused to believe she was a marketing ploy, insisting they had heard her on the radio or even seen her on TV. Besides, her portrait was printed right on the products. There had to be a real Betty Crocker most people reasoned.

By the 1960s, the portrait of Betty Crocker had changed several times to meet changing styles and fashions, but few noticed. Betty Crocker had always seemed to fit the times. Consumers and fans trusted the trademark red that she wore and her signature. For those who wrote in wanting a photo of Betty, they were politely told she doesn't like her picture taken. Those wanting a date or to marry her were politely declined.

Today customers are able to write to Betty on her Web site, www.askbetty.com, and they will receive a reply by e-mail, bearing the Betty Crocker signature.

OUT OF THIS WORLD:
THE MILKY WAY IN MINNESOTA?

- 1924 -

THE TWO MEN SAT TOGETHER, FATHER AND SON, drinking malted milks at a soda fountain in Minneapolis. Frank Mars had recently bailed his son out of jail in Chicago. The younger Mars, an enterprising, ambitious salesman named Forrest, had run afoul of Chicago city law when he'd overzealously plastered unauthorized Camel cigarette posters all over town.

Although the men hadn't spoken in years, Frank, who was beginning to taste success with candy making, helped his son out of the jam. They had little in common. Frank had divorced Forrest's mother and the young man grew up in Canada, far away from his father in Minneapolis.

By the time the men reunited, Frank was on his way to becoming a successful Twin Cities candy maker after several failed attempts at being a confectioner.

Forrest had decided to try college in California. Forrest spent more time making money by managing the University of California's cafeteria buying program than actually attending college class. Instead of spending the summer in California, Forrest hit the streets of Chicago as a cigarette salesman, looking to make even more money.

As the two men talked at the soda fountain, Forrest's natural business genius shone through. He immediately saw problems with his father's business model.

"You're making money, but we can't sell them the candy bars anywhere outside of Minnesota," Forrest recalled telling his father. Oftentimes, candy makers couldn't ship their goods without them going bad or, in the case of chocolate, melting.

When Forrest criticized his father's business, Frank replied, "Well, what would you do?"

"Put this chocolate malted drink in a candy," Forrest replied.

Later Forrest recalled that he had merely said the first thing that came into his mind. Frank took his son's advice and developed a candy bar called the Milky Way.

Presumably, the name comes from the malted milkshake that gave Forrest the idea. Despite the mediocre quality products Frank Mars used to develop the first versions of the Milky Way, the bar was a hit.

"He [Frank] put some caramel on top of it, and some chocolate around it and not very good chocolate, he was buying cheap chocolate but that damn thing sold. No advertising," the reclusive Forrest recalled in one of the only interviews he ever granted to a reporter.

Not only was the Milky Way a hit with the customers, it was also business success. The Mars candy makers were able to offer a bigger candy bar that contained much less chocolate, the most expensive ingredient in most candy. Most of the bulk came from the nougat, made primarily from eggs. More importantly, the nougat stayed fresh because it was enrobed in chocolate and it could travel farther and last longer on the shelf. The bar was a smash hit, making nearly a million dollars in the first year on the market.

The two men only spent eight years together, but during that time they laid the foundation of what was to become one of the most secretive, yet successful companies in America. Forrest left the company after a fight with his father, convinced the elder Mars was not aggressive enough with his company. Forrest moved overseas, working for rival Nestle, although the candy giant didn't know who he was. When news of his father's death reached him, Forrest nonchalantly ignored the news and continued learning the secrets of candy-making. He later came back to regain his place as an American candy tycoon, implementing the lessons he learned from his father and his father's rival. Today, the profits are estimated at $17 billion—estimated because the company is still privately held by the reclusive Mars family. The Mars candy empire would come to include such household names as Snickers, Skittles, Uncle Ben's Rice, Kal Kan, and Twix—all thanks to the success of the Milky Way bar.

THE MILFORD MINE TRAGEDY

- 1924 -

As the day shift worked hauling the manganese-rich ore out of the Milford Mine in north central Minnesota, the lights throughout the mine died. This was not necessarily unusual. A generator may have failed, or maybe a power cable had been cut. Miners lit the carbide gas lamps on their helmets and resumed their work.

But, just as soon as the gas lamps were lit, a rush of air blasted through the tunnels extinguishing them. The men tried relighting the lamps only to have the same uncharacteristic rush of wind. Wind in the tunnel was like the moon shining at noon—it just didn't happen. The air 150 to 200 feet below the surface in the mine was usually cool and still. This wind was warm. Something was wrong.

Miner Frank Hrvatin—only fourteen years old at the time—had just finished dumping a load of loose iron ore down a transfer chute when he was hit by the blast of air. He looked down the chute. What he saw horrified him. Water rushed through the level below toward him. He yelled to his partner and they began to run for safety. Forty-eight miners were working below ground. It was impossible to know how many understood what was happening. Somehow, water was flooding the mine.

Hrvatin ran with five other men up a forty-foot tunnel to the 135-foot level. From there, the men had to travel the length of almost two football fields before even getting to a ladder which might lead them to safety.

Hrvatin ran for safety while others were trapped or, worse, refused to believe they were in any sort of danger. Two miners, Valentine Cole and Minar Graves, turned back and went back to work, deciding they were not in danger. Cole and Graves literally walked back to their deaths.

"We heard that water coming down the drift," Hrvatin remembered years later in a newspaper interview. "We didn't know if we were going to make it. We just ran and ran for our lives."

Mud and water flooded the mine, literally nipping at the feet of seven men who managed to make it to the ladder that led out of the shaft. Forty-one remained trapped, some stuck in a heavy mud which anchored them like concrete. Cole and Graves realized their mistake and tried outrunning the floodwaters. Their bodies were found fifty feet from the shaft, their arms wrapped around each other, their feet anchored to their grave by mud.

The lucky seven climbed the 135 feet. As Hrvatin climbed the ladder system, an older miner, Matt Kangas, grew weaker and weaker. Hrvatin pushed Kangas and himself up the ladder even though Hrvatin was still shy of his fifteenth birthday. Meanwhile, Hrvatin's partner, Harry Hosford yelled for the men to move faster because the water and mud continued to swirl around his ankles.

Kangas had been through this before, once in Michigan.

"I knew what it was," he said to a newspaper reporter. "I knew if we lost a minute it was too late. I yelled. Then I ran like hell. We can't save our life no more if we don't run. I know. So I run. No time for the gates, no time for the cage. No time for anything. I just run, and fall down, and run some more."

When the men made it to the top of the mine on the dreary gray winter day, their legs gave out and they collapsed. Alarm sirens wailed and bells pealed. Below, the water rose to within twenty feet of the mine shaft's opening. In all, it took less than twenty minutes for the mine to flood.

News of the disaster rang throughout the town.

Jennie and Maybelle Myhres worked at the Aitkin-Deerwood Telephone Company in Crosby, Minnesota, near the mine. They were the first to receive the panicked call for help.

"Operator, get help quick! There has been a severe cave-in and the mine is flooded. Nearly all the men have been drowned," the voice on the other end said.

The two sisters knew the mine well. Their brother, Arthur, worked there, helping to support their widowed mother. Despite their fears for their brother, the operators continued to place calls for help and the calls of relatives who were trying to learn the fate of the miners who went to work that morning and wound up going to their graves.

In seven hours, four switchboard operators handled about seven thousand calls.

Huge pumps arrived nearly immediately to clear the Milford Mine of water. However, everyone knew that there would not be a rescue mission, just a recovery of the forty-one bodies below. Removing that much water would likely take weeks, maybe even months. Within a day, seven thousand family members and friends had arrived in the little mining community, seven miles from Crosby, Minnesota. Two pumps started emptying the water at a rate of ten thousand gallons per minute. Crews worked day and night to keep the huge pumps working and diverting the water to Wolford Lake, a few thousand feet away. Other mine pumps from around the state were sent to Milford by train. Often crews came with the pumps from other mines to assist with the operations. Special high voltage lines were run to feed the electric-hungry pumps.

The crews not only had to drain water, mud, and sand from the mine, but also water from Foley Lake to keep more water from rushing into the mine. The sand and the mud made the "dewatering" a difficult task. By March 28, nearly two months after the disaster, only two bodies had been recovered, those of Cole and Graves. Thirty-nine remained below ground.

It took the pumps about three months just to clear enough water to get to the bottom of the shaft. Most of the tunnels were filled with sand, debris, and mud. The veins, or tunnels, of the mine had to be cleared of the muck before any recovery could begin. Crews found not only the bodies of the workers, but also dead fish, turtles, and even muskrats. The stench was so strong that undertakers recommended covering everything in a coat of formalin solution. Those working to recover the bodies managed only fifteen or twenty minutes of work below ground before having to come up to the surface for fresh air.

The last body was recovered nine months after the disaster. Toward the end of the recovery process, most bodies were put in caskets, wrapped in gauze and covered with a glass top to seal off the stench.

"You could see a body there, but you couldn't recognize what it was," said Mike Zauhar, in the book, *The Milford Mine Disaster*. Zauhar was thirteen at the time his uncles were both recovered from the mine. "It was wrapped like a mummy."

Work to reopen the mine began almost immediately after the recovery efforts. Mine owners sunk $250,000 into it, restoring and safeguarding it against a similar occurrence. The owners also formed a special inspection committee that made frequent on-site evaluations. Despite the safety measures and a restored mine, another disaster hit—the Great Depression. With the Depression, the demand for high-grade ore dropped, and the mine closed in 1932.

From 1917 through 1932, nearly 1.3 million tons of ore were shipped from the Milford Mine. Today, the site of the mine is barely recognizable. All the buildings are gone.

Survivors didn't fare well in many cases. They lived on, but with little help from the government or the mine operators. The forty-one miners who died left behind thirty-eight widows and ninety-six children. The Red Cross provided some emergency funding. The mine company paid some compensation. State assistance for widows with families only paid up to twenty dollars per week for no longer than seven years, forcing many families into poverty or to live with relatives.

Still, the disaster didn't stop 124 miners from going back to work at the mine in 1925, nor did it stop Hrvatin, who was almost killed himself and who lost his father in the tragedy.

"I was with my kind of people—miners."

MINNESOTA'S OLDEST RESIDENTS

- 1931 -

Just outside Pelican Rapids, road crews worked on U.S. Highway 59. As they leveled the road on June 16, 1931, the grader plowed into soft earth, twelve feet below surface grade. Carl Steffen followed behind the road grader and stopped the operator when he noticed it had uncovered something shiny. He bent down to give it a closer look and saw the metallic lining of a clam shell which had been crushed by the grader. As he brushed away the dirt around it, his find became even more peculiar. He saw an eye socket of a human skull peering back at him.

Work stopped and the crew rushed to exhume the bones. As they dug, they found many bones of a skeleton. With the skeleton they also found seashells and a dagger made from a elk's horn. The crew made one other odd discovery: The skeleton had been covered by a layer of clam and mussel shells. Shocked, the road crew called the supervisor to come and see.

The next morning the crew turned the skeleton over to the

surprised road superintendent. He, in turn, took it to the University of Minnesota, where it eventually landed in the care of Albert Jenks.

Jenks, a renowned archaeologist, quickly recognized the skeleton's importance and soon the find was hailed as "The Minnesota Man," one of the oldest, if not the oldest, skeletons found in North America. It was probably ten to twenty thousand years old. Except for one problem: Minnesota Man wasn't really a man at all.

The pelvis bones indicated the man was really a young female. Jenks inspected the site and determined the body was not buried in a ritualistic way, meaning the young woman had probably died accidentally. Jenks determined that the shells that covered the bones indicated she died by drowning. The shells had just gradually accumulated on the river bed. Yet scientists knew from geological evidence that a body of water hadn't been near the area since the last glacial age, at least ten thousand years earlier.

Minnesota Man, later renamed "The Minnesota Woman," became an important find that changed how archaeologists and historians viewed human history in North America. Minnesota Woman's presence suggested her ancestors had arrived centuries earlier than many had previously thought. Minnesota Woman's ancestors had come to the continent from eastern Asia, possibly by a land bridge. They were prehistoric hunter-gatherers, arriving even before the time of pharaohs in Egypt. Minnesota Woman and her tribe would have lived in the shadow of the glaciers, hunting woolly mammoths and other land mammals.

She could have been fishing and fallen through the ice, or have fallen off a crudely made boat. Her body was then covered by mud, silt, and eventually mussels and clam shells that sifted to the bottom.

Not only did the skeleton cause historians to rewrite what they knew of North American history, but it also changed the view of primitive Native American culture. The conch shell that was found with the skeleton was probably worn as a necklace. Scientists surmised this by the holes drilled in it. But, the conch, a *Busycon perversa,* is only found in Florida.

The excitement of Minnesota Woman's discovery was still fresh when William Jensen noticed bone fragments and a spear point in the gravel of his driveway being delivered at his Browns Valley grain elevator. As luck would have it, Jensen was an amateur historian and archaeologist who rushed to the gravel pit to see if there were any more bone fragments. Jensen took an old auto license plate and using it as a makeshift shovel, poked where crews had been excavating at the bottom of the pit. Almost immediately, he struck of a brown flint knife. Jensen began to exhume the knife along with the rest of the skeleton. He then sent the skeleton to archaeologist Albert Jenks.

Jensen loaned the skeleton to Jenks on the condition that the bones be returned to him after study. Jenks took meticulous records of the skeleton, believing them to be between eight and twelve thousand years old, corroborating the earlier discovery that human civilization had existed deep in North America during the last glacial age. He termed this new find "Browns Valley Man."

Not only did the two skeletons discoveries rock the scientific world of the early 1930s, but they also raised more questions than they seemed to answer. In 1999 when researchers examined the two skeletons, they discovered that the skeletons didn't really match any modern American Indian types.

"They do not have the broad faces, they do not have the big prominent cheekbones that you think of as the more traditional features of the Chinese and American Indians," said Doug Owsley, of the Museum of Natural History, in an interview with the *Vancouver Sun.*

The skeletons more closely resemble those of people who lived in Indonesia or even Europe.

This has led some anthropologists to formulate a new opinion of how settlement on the North American continent occurred: A growing number of scientists now believe that many waves of settlers from different points of the globe arrived instead of crossing a land bridge.

While the new theory is not without its detractors, the two skeletons found in Minnesota have caused the entire world to rethink what we believe about those who lived on the continent ten thousand years ago.

Owsley said, "When you think of this time period, you imagine those people running around chasing big game, wearing skins, throwing spears, looking like today's Siberians. That probably isn't true."

SPAMTASTIC:
THE INVENTION OF SPAM

- 1936 -

JAY HORMEL WAS SICK AND TIRED of having his best ideas stolen by unscrupulous competitors. He would toil deep into the night pondering new inventions and products only to learn of competitors getting to the marketplace first or stealing his ideas. He was intent on creating something that had never been created before. In the long, cold winters of Minnesota, Hormel fleshed out his ideas, but few ever bore successful results.

As a child, Hormel was thrifty. For example, he and his schoolyard pal, Ralph Daigneau, paid women 2 cents a pound for leftover cooking grease. The boys then turned around and sold it to the soap-making division of his father's company, Hormel, for 4 cents a pound.

Hormel's father, George Hormel, required some of that creative genius to help fend off competition and increase profits for the family business, Hormel meats, based in Austin. As a father-son team, they decided to venture into the canned ham business. Beginning with the wasted meat from pork shoulders, they cubed the meat, and tested prototypes at cocktail parties.

At first the luncheon meat was sold in six-pound cans. Obviously, this was not marketed to an average family, but to delicatessens, butchers, and grocers, who would slice the loaf for customers. However, competitors soon caught on to what Hormel was doing and quickly flooded the market with their own six-pound luncheon loaves.

Hormel decided his product was good but that it needed to be sold in smaller sizes. Twelve ounces could be marketed and sold directly to customers in its own one-of-a-kind, recognizable packaging. Hormel reasoned that twelve ounces would be enough to feed a family of five for dinner with leftovers for sandwiches the next day.

Hormel also insisted that the product have a unique shape and something that differentiated it from other brands. He decided on a square shape that would allow the meat to fit neatly on bread. He sent employee Julius Zillgitt to a local grocery, where Zillgitt found a can of Mazola oil that was too large, but the right shape. Zillgitt traveled to a tin shop, where he had a worker cut the can down to the twelve-ounce size.

When the first prototypes were processed, Hormel discovered a problem. When the cans were opened, instead of finding twelve ounces of glorious pink lunch meat, they found eight ounces of hardened meat and four ounces of liquid—not exactly an attractive nor appetizing prospect for a hungry family of five.

Because of the smaller size and the type of shoulder meat, canning the new meatlike invention wasn't like canning ham, beef, or chicken. The meat had to be packaged differently so it wouldn't break down. The Hormel Company called it the battle against loose juice. Hormel scientists eventually discovered they could solve the problem by mixing the meat in a vacuum and then putting it in a can differently.

Now that Hormel finally developed the product he'd been looking for, only one more obstacle remained: What to call the pinkish blob of pork?

The company searched for an original name. "Brunch" was a popular suggestion, a term borrowed from a novel written by Minnesotan Sinclair Lewis. However, company officials let the term pass when they discovered Lewis's character in the novel dies from typhoid fever after eating brunch. The new product narrowly avoided being named "Spic," a derogatory term, which happens to be an Old English word for fat or grease. Though the name was never used, it was later dusted off and used as a shortening product. Then in a fit of either inspiration or desperation Hormel threw a New Year's Eve party, where guests received a drink for each name idea they offered.

"Along about the fourth or fifth drink they began showing some imagination," Hormel later said in an interview.

The reward for the best name was one hundred dollars. It went to Kenneth Daigneau, an actor of minor fame and the brother of Ralph Daigneau, who had collected grease with Hormel as a child. By then Ralph was a vice president at Hormel.

Company lore says the name is a combination of the words "spiced" and "ham." Others claim the word is an acronym for Spiced Pork Austin Minnesota. Legends have taken root about the luncheon meat, including that the recipe for SPAM came from Jean Vernet, a French chef working for Jay Hormel, but the namesake was never in dispute.

Again, Hormel's competitors launched products to compete with SPAM, including Treet and Prem, but none compared. Hormel was no longer frustrated; he was delighted.

Even recently the name "SPAM" stirred some controversy. For a time in the late 1990s, the Hormel company sought to put an end to the practice of calling unwanted e-mail, "spam." After a couple of years of futile legal posturing, Hormel abandoned its fight to disassociate its trademark name from unsolicited, commercial e-mail. Instead, the company simply asked that the media and others refer to SPAM in all capital letters.

Jay Hormel himself would have been amazed by SPAM's enduring popularity. A SPAM museum, decked in the trademark blue and gold, sits right off Interstate 90 in Austin. Hormel Foods Corporation produces SPAM at the rate of forty-four thousand cans an hour. That works out to about seventeen thousand pigs per day just to make SPAM.

And, SPAM is a worldwide hit.

The pink luncheon meat is even more popular in Asia than in America. In Japan SPAM is given as a gift. In the Philippines, Hormel has upscale restaurants in malls which feature menus teeming with SPAM items, including SPAM spring rolls and SPAM salad with sesame dressing. And, Hormel supplies SPAM to restaurants like Burger King in Japan and McDonald's in Hawaii. As Hormel's slogan once said, "Love at first bite."

THE ARMISTICE DAY BLIZZARD

- 1940 -

THE WEATHER COOPERATED PERFECTLY for duck hunters along the Mississippi River as a gun-metal gray sky hung over the forests and a light rain fell. Ducks hovered in the air and slowly glided to the ground and nearby waters. On November 11, 1940, the weather in Winona at daybreak was fifty-two degrees, almost too warm for duck hunting.

At the same time, far, far away near Kansas City, Missouri, a freakishly low-pressure system pushed through the Midwest, causing the gates of a wintery fury to fly open. The weather system moved so quickly that forecasters and broadcasters barely had enough time to warn anyone. Nine hours later, in the afternoon, most of Minnesota's temperatures plummeted to the mid-teens and everything was wrapped in blankets of snow and ice.

Dan Kukowski was duck hunting in a canoe in the Winona area when he noticed an increase of wind around 11:00 a.m. He also couldn't help but notice ducks seemed to be coming from everywhere. What he didn't realize was that the ducks were trying to escape the violent weather that was pursuing them.

An hour later Kukowski estimated the temperature had fallen to around ten degrees on the water. But he didn't believe the bad weather would last too long, and he remained on the river, hoping it would subside. Before 1:00 that afternoon, when the willow trees were laying flat, Kukowski decided the weather wasn't going to improve and he crossed the river in his canoe, leaving his hunting partner, John Zenk behind. It was so cold that Kukowski's shotgun had frozen. Winds were clocked at nearly sixty-seven miles per hour.

"[Zenk] was under the raincoat hugging the dog. That's what kept him alive. The dog saved him," Kukowski recalled years later when speaking to the *Star Tribune*.

Survival for Zenk meant hugging his dog tightly for hours, for others it meant taking refuge under a boat, even burning expensive wooden decoys to stay warm in the freezing hell.

Duck hunters weren't the only ones who got caught in the Armistice Day Blizzard. A passenger train near Watkins collided with a freight train during white-out conditions. The passenger train didn't see the signal that set it on a collision course with a freight train. Two crew members were killed as the trains collided. Watkins residents had to form a human chain to lead the passengers to safety because visibility was zero during the snowstorm. As they were rescuing the passengers, one of the train's steam whistles howled. It had been stuck open in the collision and it continued howling through the night until the steam ran out.

"A low mournful sound like it was crying over the accident that happened," said Wendelin Beckers, who was twenty at the time and heard the thunderous crash. He recounted the story in 2000 in a Minnesota Public Radio interview, "Those steam engines were smashed right head on together."

Those who survived the storm told stories of how they stayed alive; some who died left evidence of trying to brave the unexpected cold.

One storm survivor recalls trying to use the gasoline from a boat engine to kindle a fire only to have it dissolve in the air as he tried to dump it out. Some tried to find shelter on the many islands in the middle of the Mississippi River. They turned boats upside down and tried to kindle a fire with oars or decoys. Others would shadowbox to keep warm and fight off sleep. Still others marched in place, trying to keep warm and stay awake. Several reports told of dead hunters with bruises on their chests from where they'd beaten themselves to try and stay awake. Some had fired their guns in an attempt to draw the attention of imagined rescuers while still others burned the wooden butts of their guns.

Gerald Tarras lived to recount how the sudden storm that took the lives of his father, brother, and friend. He managed to escape only with frost-blistered hands.

Tarras's day began at 10:00 a.m. when the foursome went out duck hunting on the Mississippi River. By then the rain began to pick up and the temperature had dropped nearly thirty degrees. By noon the temperature had fallen even further and the wind howled. Tarras's father decided to turn around, but the wind and snow were too fierce and it was impossible. The four huddled together with their black Labrador dog. Slowly, one by one, each of the men died. First, Tarras's friend, then his brother, and then his father. Tarras was left alone until the storm cleared and a rescue pilot spotted the boy from the air.

Even trying to rescue the frostbitten survivors or find the dead was treacherous. Max Conrad, a legendary Minnesota aviator, flew rescue missions over the Mississippi River looking for survivors. He fired up his Piper Cub, which took ten men to hold it steady against the raging wind as he readied for takeoff.

Conrad spotted several stranded hunters. When he found survivors, he dropped a package full of food, cigarettes, matches, and a pint of whiskey. He flew low, slowed a little, killed the engine, and yelled, "Hang on, help is coming." A government tug followed Conrad's plane as he found survivors from the air.

The city of Winona turned its city garage into a morgue, where bodies were thawed for identification.

The next morning the wind and snow subsided, but the temperatures continued to hover in the low teens, even into the single digits. The cold had frozen the river and some hunters simply walked to safety.

The storm left an amazing amount of snow. The central pressure, which had worked its way from the Texas panhandle north, had been down around twenty-nine inches of mercury, said meteorologist Paul Douglas, who has studied the storm in-depth.

"It doesn't get much stronger than that. It allowed moisture from the south to interact with this fresh Canadian air mass to the north," Douglas said in a radio interview in 2000, "and those two converged to produce this incredible intensification to the point where you really did have what you could call an inland hurricane."

The storm covered nearly a thousand miles, causing death and destruction. The Twin Cities had sixteen inches of snow, while Collegeville had twenty-seven inches. Winona had twenty-two inches of snow, and winds were estimated at between fifty and eighty miles per hour, which caused waves on the Mississippi River of almost six feet.

Turkey farmers across the state were hit hard. Nearly one million turkeys died, less than two weeks before Thanksgiving.

By the end of the one-day ordeal, forty-nine Minnesotans were dead and scores of others were treated at hospitals throughout the state for frostbite and other injuries. Across the Midwest, over 150 died in the blizzard, sixty-nine people alone died on boats or ships on Lake Michigan.

Just recently the Armistice Day Blizzard made headlines again. In 1999 fifty-nine years after the blizzard, the city of St. Cloud, Minnesota, honored Marshall Jones for his bravery during the storm.

He was driving home on November 11, when his car got stuck in traffic because of the storm. He made his way on foot to a nearby high school. There he noticed a teacher who didn't want to wait out the storm. She walked out of the building with just a coat and hat. Jones knew she wasn't prepared for the intense storm. He asked the high school students which way the teacher usually walked home and then quickly left to find her. As he made his way toward Lake George, he saw a shadow of a person standing still. It was Gertrude Gove, the teacher. He tried to get her to follow him, but she just stood there in shock.

"She gave up," Jones said in an interview fifty-nine years later.

So, he had to snap her out of the shock.

"I had to cuss at her. I let her have it," Jones said.

That snapped Gove into moving and they both took cover in a nearby store.

Even though Jones faced other challenges in his life, including being a wild animal handler, a police investigator, and a bus driver, he said his hands still went numb every time he thought about the Armistice Day Blizzard.

THE NIGHT THAT MUSIC LIVED

- 1959 -

"Hey, what's your name?" the emcee asked the nervous fifteen-year-old bandleader waiting to go on stage at the armory in Moorhead, Minnesota.

The young man, Robert Velline, froze.

The emcee stared at him. Velline then said the first thing that came into his head.

"The Shadows."

The truth was his band didn't have a name, and the day before Velline had been just another kid ready to see his rock 'n' roll idol, Buddy Holly. The name "the Shadows" stuck, and Velline would later be known to the rock 'n' roll world as Bobby Vee.

The day before, now known as the "Day That Music Died," was the day that three well-known rock stars, Buddy Holly, Ritchie Valens, and the Big Bopper (J. P. Richardson), lost their lives in a plane crash in an Iowa cornfield.

Ron Lucier, a Moorhead disc jockey, had booked Holly, Valens, and the Big Bopper to perform in Moorhead for the Winter Dance

Party. Now the show was missing its three biggest stars. Lucier and the promoters hoped the show would continue and spoke to acts like Dion and the Belmonts, along with Holly's bassist, Waylon Jennings, about performing. Calls poured into local radio stations with fans begging for the show to go on.

Charlie Boone, another local disc jockey, put out a call for other acts that might be able to help with the show. Robert Velline happened to hear the call on KFGO, a radio station in Fargo, North Dakota. Velline, a young Fargo Central High School student, had formed his band just two weeks before, and the new band hadn't played in public yet. The unnamed band called the radio station and was told to show up early because they would be the opening act.

Velline and his brother, Bill, had been waiting for a moment like this for awhile—they just didn't know it would happen so soon. They had practiced their own versions of popular songs in the living room of their house.

Having never played in public before, the band had to find outfits. They stopped by the local JCPenney store for pants and then bought sweaters from a thrift store. They arrived in Moorhead, and before the band knew it, they were being thrust on stage, hailed as the Shadows.

"The fear didn't hit me until the spotlight came on, and then I was just shattered by it," Velline recalls on his Web site. "I didn't think I'd be able to sing. If I opened my mouth, I wasn't sure anything would come out."

Lucier's concert turned out to be part concert, part wake. More than two thousand people jammed into the Moorhead Armory, which technically could only hold seventeen hundred. More were lined up outside. So many people came because of the national publicity the concert garnered in the wake of the deaths of Holly, Valens, and the Big Bopper.

Charlie Boone emceed the show and recalled in the book, *The Day the Music Died*, "I don't know if people came to the armory expecting to see coffins laid out in front, but there was a curiosity factor."

Before the Shadows went on stage, Frankie Sardo, a minor celebrity on the tour, did a tribute to the performers, including Valens's hit, "Donna."

When the Shadows took the stage, Robert and Bill sang tunes by the Everly Brothers, Ronnie Hawkins, and Little Richard. It wasn't the cover songs that made Robert and his band famous that night; it wasn't even that they were thrust into the national spotlight. It was the way Robert sounded. The familiar hiccup and drawl he used—the style he copied from his idol, Buddy Holly—stunned the already shocked crowd.

Burnell Bengtsson saw the Shadows save the maudlin concert.

"When [Robert] sang, he sounded like Buddy Holly. Everybody gasped. It was a really strange feeling," Boone said in *The Day the Music Died*.

Bengtsson was impressed with the Shadows.

"Good job, boys," he said. "Call me if you're looking for work."

The next day Bengtsson became the agent for the Shadows, and he booked the group two weeks later for a gig forty-five miles away from Fargo. The Shadows drove to the performance in a station wagon in temperatures that hovered near zero, and the station wagon didn't have heat. At one point during the show, the benches that had been put together to form a makeshift stage pulled apart, causing the amplifiers to tumble to the floor. The band earned sixty dollars for the performance. From there the band went to Minneapolis, where it recorded "Suzie Baby," a Buddy Holly sounding single, which appeared on the Top 100 pop chart. The song's success, including a number-one position on the local Minneapolis pop chart, earned Robert a recording contract with Liberty Records.

But before the Shadows went back to Fargo and Bobby Vee went on to become a well-known 1960s pop star, the band had one more encounter with rock 'n' roll history.

During the summer of 1959, as the Shadows toured the Midwest, they looked for a rock 'n' roll piano player like Little Richard or Jerry Lee Lewis. Bill Velline met a spindly, skinny guy at Recordland in Fargo who claimed to have toured with Conway Twitty. The piano player's name was Elston Gunnn.

The first gig with Gunnn on the piano was in Gwinner, North Dakota, where an old, out-of-tune piano almost drove the awkward piano player crazy. The next night the Shadows played a concert where the only piano was also out of tune. The quirky Gunnn had about enough of the poor pianos, but managed to part amicably with the Shadows.

Gunnn's real name was Robert Zimmerman. He was a kid from the Iron Range of Minnesota, who later became known to the world of music as Bob Dylan.

Dylan later wrote about Bobby Vee in his autobiography: "We had the same musical history and came from the same place at the same point of time . . . I'd always thought of him as a brother."

IT'S A HIT: THE HOMER HANKY

- 1987 -

TWENTY-TWO YEARS HAD PASSED since the last time the World Series had come to the land of ten thousand lakes. In 1965 a stunned crowd watched legendary southpaw Sandy Koufax pitch a 2–0 shutout on just two days rest. The Twins went down in game seven to the Los Angeles Dodgers.

In 1987 the Twins brought the World Series back to Minnesota, and fans had plenty to cheer about. One of the metropolitan newspapers, the *Star Tribune,* wanted a piece of the marketing action. Marketing manager Terrie Robbins convinced upper management at the newspaper to print sixty thousand handkerchieves to give away at the first home game of the playoffs.

The idea was quintessentially Minnesotan, where the fans were too polite for air horns and the weather was generally too chilly for body paint, even inside the climate-controlled Metrodome.

Fans stood in line for up to six hours to buy the simple fabric hanky, a little smaller than a dish towel. The *Star Tribune* was forced to limit five hankies to a person, then two. They were a big hit. Four-teen printers nationwide worked around the clock trying to keep up with demand. Robbins said unmarked vans delivered the wildly pop-ular hankies to the *Tribune,* where they were stored in the vault along-side the payroll checks. Over one million Homer Hankies were printed in 1987—enough for a quarter of the entire state's population at the time.

Despite the success of the hanky, it looked as if heartbreak was coming soon for the Twins. They dropped a few games to the Cardi-nals in St. Louis, and the series was back in Minnesota for game six. By the fifth inning of game six, the Cardinals had a 5–2 lead and were looking to win their fourth straight game and the series title.

However, in the Twins half of the fifth, Hall of Famer Kirby Puckett led off with a single and eventually came around to score, thanks to a Gary Gaetti double. During that inning, with one Twin on base, Don Baylor came to the plate. Baylor, a seasoned veteran, had been traded to the Twins for his offensive power. He didn't dis-appoint and promptly delivered a two-run homer into the left-field seats, where thousands of fans and thousands of Homer Hankies went wild.

The scene was a sea of white. The entire Metrodome, complete with its pale white ceiling, had the look of a fine Minnesota snow-storm, and the cheering of the fans was a deafening roar. The Twins had tied the game.

A Steve Lombardozzi single pushed the Twins ahead. The Twins didn't look back from that point as Kent Hrbek put the game far out of the Cardinals' reach with a grand slam in the sixth inning.

The Twins won game six forcing a decisive game seven for the World Series.

Frank Viola—known to most Twins fans as "Frankie V"—started the final game. Though the Cardinals took the early lead, the Twins came back and tied the game in the fifth at two runs apiece. Greg Gagne's infield single pushed the Twins ahead. An insurance run in the eighth gave the Twins a 4–2 lead. Closing pitcher Jeff Reardon was brought on to finish the top of the ninth. Cardinal Willie McGee grounded out to Kent Hrbek at first for the last out.

The stadium erupted in a sea of white. The Twins became the first team to win a World Series without winning on the road. It became the first team to bring a World Series title to Minnesota, and it was the first time the franchise won since 1924, when the team was the Washington Senators and their ace was Walter "the Big Train" Johnson.

About thirty thousand fans stayed to hear Twins players talk about what the series meant to them. Twin Cities native Kent Hrbek told the crowd, "Ever since I was a little boy, I've dreamed of driving down Hennepin Avenue in a victory parade. And, I get to do it."

Sports journalist Thomas Boswell described the celebration: "Long after the game, perhaps 30,000 or more fans still stayed in the Thunderdome stands, waving their Homer Hankies, or perhaps wiping away a tear with them."

But before the Twins even won the World Series, the Homer Hanky had already become part of Minnesota history, and on October 21, 1987, the Minnesota Historical Society placed at least one Homer Hanky into its collection.

The popularity of Homer Hanky didn't end with the 1987 World Series.

The World Series came back to Minnesota four years later. Although some of the players had changed, Minnesota fans were still just as crazy about their team.

Midway through the 1991 season, sports journalist Jim Klobuchar of the *Star Tribune* reminisced, "What validated the World Series experience for thousands of people four years ago was not seven games and Frankie V., as much as T-shirts, sweaters and Homer Hankies."

As soon as tickets went on sale, Homer Hankies were in demand. At first the Minnesota Twins thought of producing their own version of the *Star Tribune*'s successful hanky. They concocted a piece of fabric that looked suspiciously like the Homer Hanky and dubbed it the "Rally Rag." Fortunately, the idea was short–lived, and the Twins and the *Star Tribune* worked out a deal to revive the Homer Hanky.

The lines to purchase the hankies were three blocks long. The newspaper gave out coffee and hot chocolate and brought in portable toilets and ropes, the kind used by Disneyland to organize long lines. Despite a maximum order of ten—then five, then two—folks brought their dogs along in attempts to thwart the rules. Fans standing in line would claim the maximum number for themselves and the maximum number for their pets—or whoever else they brought along. Robbins remembers one woman going through the line three times with a baby. When staff finally recognized her and asked to see her baby, she panicked and ran off, worrying they might try to take back the coveted cloth.

The newspaper even paid for rights to use the Tommy James and the Shondells song, "Hanky Panky." The lyrics were altered slightly to "My baby loves the Homer Hanky," and it was a hit, with entire crowds happily singing along.

Many didn't need to buy new Homer Hankies since they had the old ones from 1987. Many had to dig out their scrapbooks or even take their Homer Hankies out of picture frames to use them again.

History seemed to be repeating itself.

The series was held at the Metrodome, and the Twins entered game six to win or they would lose the series to the Atlanta Braves. The Twins took an early 2–0 lead, thanks in part to Kirby Puckett's triple. Later Puckett helped save the game with a phenomenal back-handed catch of a deep fly ball by Ron Gant to center field. The long drive appeared to be a home run until Puckett scaled the Plexiglass wall, leapt, and snared the ball, saving what would have been a two-run homer. But that wasn't the most spectacular part of Puckett's night.

With the game tied at three, Puckett came to bat in the eleventh inning and faced veteran lefty Charlie Leibrandt. With the count at 2–1, Puckett drilled the ball into the left center-field seats for a game-winning home run.

As the ball flew out of the park, thousands of fans were already on their feet, waving their Homer Hankies. The win forced another decisive game seven for the Twins.

This time it would be St. Paul native Jack Morris who would prove to be the dominant pitcher. Game seven remained scoreless through nine innings. In the Braves' half of the ninth and tenth innings, Morris was able to get through the order in just eight pitches each inning.

The Twins used their "small-ball" approach again in the tenth, with the game running late and the entire season on the line. Out-fielder Dan Gladden led off for the Twins with a broken-bat bloop single that fell into left field. Gladden, seizing the opportunity, stretched the tiny single into a double. Second baseman Chuck Knoblauch sacrificed Gladden to third. The Braves walked the next two batters—Hrbek and Puckett—to intentionally load the bases. The Atlanta infield, in desperate need of a double play to get out of the inning, brought the infield in to induce the weak-hitting Gene Larkin into hitting a ground ball.

But Larkin, who hadn't seen much playing time in the field, had been watching plenty from the bench. He noticed that Braves reliever Alejandro Pena had started hitters off with a fastball. Larkin approached the plate, waited on the fastball, and nailed a fly ball into left field. He instantly knew it was deep enough for a sacrifice fly, allowing them to score. He started waving his hands in excitement. The fly flew beyond Atlanta outfielder Brian Hunter, and Gladden jogged home to another World Series title.

The trendy talisman seemed to work its magic again.

THE NINTH WONDER OF THE WORLD

- 1992 -

PEOPLE CAMPED OUT JUST TO GET A SPOT IN LINE. The line wasn't for tickets to a blockbuster movie or a concert. Still, in one twenty-four-hour period, nearly 150,000 people came.

They had come to witness history and maybe leave with a pair of jeans, too. They had come to see the opening of the "ninth wonder of the world." That's what Nader Ghermezian called it. Ghermezian's family was one of the developers of the Mall of America.

The Mall of America isn't the biggest mall on the continent; that honor belongs to the West Edmonton Mall, in Edmonton, Alberta, Canada. It also isn't even the mall with the most retail space; that distinction goes to the Del Amo Fashion Center in Torrance, California. Developers dubbed it "the largest retail and entertainment" center in America, terminology fuzzy enough for no one to dispute.

Still, no matter what developers were calling the mall, it impressed those who had waited to shop until they dropped. For example, if a person spent ten minutes inside every store in the Mall of America, it would take eighty-six hours to visit every store.

Some guests were treated to a black-tie and sneakers gala the night before the mall's opening. Peter Graves, the gravel-voiced actor who had been a fixture on the television show *Mission Impossible,* opened the mall, serving as the emcee. Ray Charles, decked out in a red satin jacket, performed "America the Beautiful," and then left the stage. He was paid fifty thousand dollars to sing one song.

When shoppers walked through the doors that first day, they found 330 new stores waiting for them. For ten thousand employees, it was the first day of work in the new megasurroundings. Within the first four months of opening, one poll estimated that over half the population of the Twin Cities metro area had visited the Mall of America.

The mall was so big that it printed maps to help shoppers. It also offered beepers for rent to those worried about getting lost or needing to stay in touch. The mall itself had three miles of corridors and cost about $625 million to build. The mall also boasted its own postal zip code and an eighteen-hole miniature golf course. In addition to providing jobs for ten thousand Minnesotans, the mall also had its own school for employees' children. The Mall of America was even equipped with family rooms that had TVs and microwave ovens so that families could "decompress" and then return to shopping, of course.

Teenagers, the most exacting and frequent customers of malls across America, were impressed. "The smaller malls are getting boring. This is exactly what we've been waiting for," said Eileen Brandel in a *Star Tribune* article. Brandel was fifteen when the doors opened. She was one of the 150,000 who came through the doors the first day, and she brought five friends with her. She arrived at 6:00 a.m., just to get a good spot in line.

The choices shoppers had inside the mall seemed to be as vast as the structure itself. For example, inside Dayton's, one of the mall's four anchor stores, sat one hundred thousand pairs of shoes and sixty-six different sizes of men's shirts. Business also seemed brisk. For instance, Oshman's sports store had forty-two cash registers operating and it reported selling one pair of in-line skates per minute.

The Mall of America also had an impressive parking lot. Two parking ramp behemoths, able to hold thirteen thousand cars, had been constructed. These structures were the largest parking ramps in the world, and cost almost six thousand dollars per space to build. On opening day, the parking ramps had to close for nearly an hour around noon because they had filled completely.

The four highways surrounding the mall were also jammed on opening day, and those who didn't want to brave the crowded parking ramps could park off-site. Mall and city officials, expecting a huge turnout, had arranged for another eleven thousand parking spaces around the area. Buses also carried over four thousand people to the mall's doors.

In the first half day of operation, the mall checked out three hundred strollers, twenty-five electric carts, and seventy-five wheelchairs. Mall workers handled fifteen hundred calls for help and fifteen children were reported lost or missing, all of whom were safely returned to their parents.

Shoppers who didn't want to lug all of their purchases around while they continued to shop had the option of forwarding packages to a valet-style service at the parking ramp until they were done shopping.

The Mall of America also owns the world record for the most indoor plantlife. Thousands of trees and shrubs give the air-conditioned facility a more natural feel. And, to avoid the use of chemical pesticides, mall management uses pest-eating ladybugs. And, even in Minnesota, a state known for its frigid winters, the mall runs air-conditioning year-round because of the massive amounts of heat given off by the over one hundred thousand people who visit daily.

Like the West Edmonton Mall, the Mall of America featured an entire amusement park contained completely inside the mall. The *Peanuts* comic strip–themed amusement park is small peanuts for a building that could hold seven Yankee stadiums inside. Camp Snoopy took up seven acres and was adapted to be indoors. For example, the roller coaster was outfitted with silicone wheels so that shoppers wouldn't have to yell just to have a conversation. On opening day, lines of people snaked around the park. Twenty thousand rides were taken in the first four hours; most people had to wait forty-five minutes for a ride. To alleviate the lines, Camp Snoopy employees walked through the crowd selling tickets to rides.

Lines also formed for the restrooms around the mall on opening day, and the average wait was about ten minutes. Mall officials had a plumber on-duty throughout the day, just in case.

In 2002 mall officials estimated that at any given moment during a normal business day, 125,000 people were inside the mall. Mall security officials also estimated that at least a dozen of those people each day are committing crimes, mostly shoplifting (30 percent of all crime in Bloomington, Minnesota, occurs at the Mall of America). The mall employs 130 security officers and has a police substation of five officers. Since the mall opened in 1992, it has also witnessed over a half-dozen stabbings and a few shootings.

The Mall of America also estimates that it draws 40 million people annually through its doors, more than one-and-a-half times as many people as Disneyland. While the mall's attendance and its over five hundred stores are impressive, its financial numbers are less impressive. For example in 2002, South Coast Plaza in southern California brought in almost $200 million more than the Minnesota megamall, despite having two hundred fewer stores.

The Mall of America continues to set shopping trends, though. Recently, in an effort to capitalize on its proximity to the Twin Cities airport and many area hotels, the mall has been advertising to travelers to take advantage of Minnesota's no sales tax on clothing. The number of stores in the mall has grown, and some of the attractions have changed since the opening. For instance, there's now a 1.2 million-gallon aquarium, a church, a college, and the world's largest Lego play area.

"The Mall of America was never a place to go and shop for things. It was a place to shop for an experience," said Therese Byrne, a retail and real estate strategist on the mall's tenth anniversary.

THE RAGIN' RED:
THE FLOOD OF EAST GRAND FORKS

- 1997 -

THE OILY SMELL OF A DIESEL ENGINE hard at work and the chill of winter were in the air. Everywhere the few remaining residents in East Grand Forks looked, there was muddy river water. John Zavoral had been working around the clock building earthen dikes to protect the city. He noticed that the few residents who remained seemed to just wander around. He likened them to zombies, wandering aimlessly, helpless against the rising waters.

The Red River had flooded before, but officials had always told the city it would be safe. The floodwaters that were puncturing the dikes and lapping over the fortified walls of sand actually began months earlier.

Winter began right on time in November 1996 when a blizzard dumped over a foot of snow on the area. Seven other blizzards, including one in April 1997, which dumped more than six inches of powder on the area, made for a high snowpack of a hundred inches. That final blizzard, named Hannah, started out as rain and turned to ice when winds of over forty miles per hour caused power lines to break, leaving nearly three hundred thousand people without electricity.

Residents had grown tired of the treacherous weather. The winter was so severe that Minnesota governor Arne Carlson classified the region a disaster area because of the freezing temperatures and the late snowfall. By the end of winter, East Grand Forks children had missed eleven days of school. Even the children cringed at the idea of another snow day.

Most of the residents had been so busy shoveling snow and warding off blizzards they hardly noticed the visits from the Federal Emergency Management Agency (FEMA) and its pleas for residents along the Red River to sign up for flood insurance. FEMA was getting worried about all the snow piling up along the Red River Valley. FEMA ran ads showing people hanging in trees as water raged around them. And, even fewer residents seemed to notice that the government cut the sign-up time in half for the National Flood Insurance Program.

On February 13, the National Weather Service warned of a catastrophic flood. Despite even that warning, by the beginning of March virtually no one had enrolled in flood insurance.

Soon the river was full. It began when the snow pack near Breckenridge melted quickly. When flooding takes place in the Breckenridge area, it generally takes twelve days for the water to reach the Grand Forks area. But because of the heavy snow, the river crested there twice in 1997, once on April 5 and then again on April 14.

In Ada flooding was so extreme that a herd of cattle died in a field as the rising waters froze overnight. Vice President Al Gore visited Breckenridge and neighboring Wahpeton, North Dakota, to survey the cities, which were devastated on April 13 when the levees surrounding the city broke. The area was declared a federal disaster area. Still, residents upriver believed they had escaped the flooding.

The Red River of the north is an anomaly because it flows northward to Canada, instead of south, tracing the border of North Dakota and Minnesota. Unlike the Mississippi River, which separates part of Minnesota from Wisconsin, the area around the Red River consists of flat fields, instead of bluffs and canyons.

The river itself is not particularly deep, which means flooding is likely; and when flooding happens, the water spreads out over the flat land. Most rivers flow south, so when snow from the north starts to melt, it typically flows to warmer regions. When snow melts in the Red River basin, the resulting water flows to the north. However, a problem arises when water starts flowing to colder regions still covered in ice: This creates a dam, called an ice jam. When an ice jam forms in the river, the river floods.

Water began rushing from Breckenridge and Wahpeton north toward Grand Forks at the same time the Hannah blizzard blanketed the region in snow and ice. Hannah knocked out power and destroyed weather-monitoring equipment along the Red River, crippling the National Weather Service's ability to gather information about the rising Red, just when it needed it most.

Maybe even worse, the late blizzard doubled the amount of snow on the ground. Forecasters realized that twice as much snow lay on the ground around Grand Forks than in the Blizzard of 1897—the blizzard by which all others were measured. More snow meant only one thing, more water.

Normally, the mean discharge of the Red River during the year is 2,630 cubic feet per second (cfs). During 1897, the number spiked at 85,000 cfs. But on April 18 the Red peaked at 136,900 cfs, the equivalent of fifty-two Red Rivers trying to make it through a single channel.

Residents rushed to fill sandbags and prepare dikes. But the water seemed to be coming from every direction, and on April 18 the dikes started to give way.

"We pretty much knew at that time no matter if we plug the holes . . . we were pretty much done for," said Zavoral in the book, *A Small Town's War*. Zavoral had been a key part of helping safeguard the city of East Grand Forks.

City officials could only watch in despair as the town was slowly engulfed in muddy water. For days the Emergency Operations Center in East Grand Forks had been a hub of nonstop activity. Machines, sandbags, emergency radios, and the news reports made for a cacophony of sounds. Adrenaline, and the desire to save the city, kept crews going around the clock.

"Everything was going and going," said Marilyn Bren, who remembered the event several years later in *A Small Town's War*. Bren was working at the Emergency Operations Center then: "We stood around and we cried. I think about it now and get goose bumps."

At 6:00 a.m. on April 19, with much of the battle lost, the Grand Forks River gauge was swallowed by the river, a single, probably insignificant act since the town was now gone, but nonetheless a fitting one that seemed to symbolize what the river was doing to an entire community.

The city of East Grand Forks was submerged and about half of Grand Forks was underwater. A fire broke out in a downtown Grand Forks building and at least four feet of water covered the streets. Fire hydrants were useless because there was no pressure in the hydrants and no fire trucks could make it through the water. Firefighters tried to battle the blaze, but without water pressure, and standing in hip waders, it was impossible to stop the inferno from the ground.

The city called for an aerial attack. With an airborne army of tankers and helicopters dropping 120,000 gallons of fire retardant, the blaze gradually subsided. Residents were stunned that in the midst of a flood, their city teetered on the verge of burning to the ground. Eleven buildings had been burned or destroyed in the blaze. One city council member who was also a veteran said Grand Forks reminded him of Vietnam.

The next day, as the sun rose on Grand Forks, the Red River continued to rise. By early that morning, Sunday, April 20, the water supply ran out and Grand Forks mayor Pat Owens ordered a twenty-four-hour curfew. Patients from the local hospital were transferred to other hospitals in the region, some as far away as Iowa.

East Grand Forks residents needed to get out of the city, but couldn't cross the swollen river. The bridge between Grand Forks and East Grand Forks was submerged. Instead, the residents traveled to Crookston, twenty-five miles away. Almost nine thousand East Grand Forks residents evacuated. Across the river in Grand Forks, 90 percent of the fifty-two thousand people left. It was the largest evacuation since Atlanta evacuated during the Civil War.

The river crested on April 21 at 54.11 feet, breaking the 1979 record by nearly six feet. The river stayed at that level for nearly a day before receding. No one had expected that much water. Just one week earlier, the National Weather Service had forecasted it would not even hit fifty feet, which lulled some residents into believing they were safe.

"They [the National Weather Service] missed it, and they not only missed it, they blew it big," East Grand Forks Mayor Lynn Stauss told national media just days after his city was wiped out.

But the National Weather Service wasn't responsible for the horrible flooding that caused $2 billion in damage to the region. A series of blizzards and storms had turned this into the perfect flood.

Much of the rural economy sustained heavy loss. Livestock was marooned on high ground, and carcasses of all kinds of animals clogged streams and tributaries. The National Guard cleaned up part of the mess as the river slowly went back down. Within a month of the flood, nearly fourteen thousand carcasses—weighing eleven million pounds—were removed.

The *Grand Forks Herald,* which would win a Pulitzer Prize for public service, continued to publish, even though its offices had been wiped out in the fire and flood. From trailers miles away, staff worked on putting the paper together. It was then printed in St. Paul, Minnesota, at the *Pioneer Press,* another Knight-Ridder newspaper. The papers were shipped to Grand Forks on an early morning flight.

When residents came back to survey the damage, the water remained shut off. Human waste and other contamination made the dirty water dangerous. It would be days before the water plant was up and running again. Meanwhile, the National Guard brought in large tanker trucks which held two thousand gallons of water and parked in a motel parking lot. People waited in long lines to fill plastic jugs and begin the recovery process.

The cleanup was nothing short of massive. Houses, buildings, and whole neighborhoods had to be completely rebuilt, not unlike the rebuilding that was needed in New Orleans after the destruction caused by Hurricane Katrina in 2005. Residents hauled 224 million tons of garbage to area landfills, the equivalent of nine months worth of garbage.

THE BODY BECOMES THE GOVERNOR

Jesse Ventura had always been a novelty act. From the pink boas and outlandish comments as a professional wrestler, to the cameo role in an Arnold Schwarzenegger movie where he famously quipped, "I ain't got time to bleed," Ventura was part novelty, part act.

That's why political scientists and some voters can be excused for ignoring the cigar-chompin' talk-show host when he announced he was running for governor of Minnesota.

The idea wasn't totally outlandish—Ventura had been elected mayor of Brooklyn Park, a Twin Cities suburb. But a peaceful suburban life and a platform of wetland preservation wasn't the same as being the governor and wrangling with the legislature.

Ventura announced that he was signing on to Ross Perot's Reform Party, but when the Texas billionaire's organization refused to cough up money, Ventura had no choice but to go it alone as he ran for the governor's seat.

By the time the governor's race of 1998 rolled around, Ventura was adamant about changing Minnesota state politics, even though he was a bit fuzzy about what he wanted to change or how he was going to change it. However, his plainspoken, sometimes blunt, but oftentimes charming wit had voters enamored with the former Navy Seal, who cast an impressive shadow at six feet, four inches.

When Ventura attempted to borrow money to fund his campaign, most banks refused to loan him money. Even though Ventura was almost guaranteed five percent of the vote and therefore qualified for state funding, he was forced to personally guarantee the credit. Finally, one smaller bank, Franklin Bank in Minneapolis, made the loan.

"He's the type of small-business borrower you want across the table. He speaks in plain English," Jim Shadko of Franklin Bank told the *Star Tribune* just days after the Ventura's election as the media scrambled to figure out how the ersatz wrestler won. "We said, 'Where's the risk?' He said, 'If there's some way I don't get five percent of the vote, the three of us (Ventura, his treasurer and campaign manager) will somehow raise the money and pay you back.'"

Then Ventura set out on the campaign trail.

Most pollsters gave him 10 percent of the vote, acknowledging that he did have a certain novelty appeal. Most predicted Hubert H. Humphrey III, the son of the former senator and current Minnesota attorney general, would easily beat the erstwhile St. Paul mayor Norm Coleman, with Ventura trailing leagues behind.

Ventura was not invited to participate in the televised debates between Humphrey and Coleman. But Humphrey insisted that Ventura be included. Many political scientists thought that the fiscally conservative Ventura would pull votes away from Coleman. The move backfired, and while the two career politicians gave politically long-winded answers, Ventura spoke plainly, even if simplistically during the debates. Sometimes his answers were impractical if not impossible to implement in state government. For example, in one of his early interviews, he vowed to eliminate the county assessor's position from every one of Minnesota's eighty-seven counties if he was elected. He claimed property taxes, which the assessor collected, were simply too high and that Minnesota was in the grips of communism.

Yet, Ventura was no lightweight in wrestling or in the political arena. He actually got cheers from a college crowd when he proposed to do away with tuition assistance, urging the students to pay for their own education. As Humphrey's campaign manager pointed out, Ventura was able to say things that would have gotten the other two candidates disowned by the voters and their parties.

"We had to entice the disenchanted voter back," Ventura told reporters after the election.

The voters had warmed to the Ventura slogan, "Retaliate in 98." They liked his commercials, which featured a Ventura action figure fighting off the Evil Special Interest Man. He also took the song "Shaft" as his theme. And, the voters chuckled when he fashioned himself after the famous sculpture, *The Thinker.*

Before long the election polls showed that Ventura had actually lured disenchanted voters back, picking up many who had never voted before. It probably wasn't surprising to Ventura, who had only voted four times himself in the previous fourteen elections. On Election Day Ventura showed up to vote in a leather jacket with fringe. According to his biography, *Inside the Ropes with Jesse Ventura,* he conducted a few interviews, but was in a rush.

"We'll do them quick," he said. "I want to see *Young and Restless.* It's my favorite TV show. I feel a lot more rested today and very relaxed. I mean, what the heck, there's no reason to get tense now. It's up to the voters and the voters will make a choice and life will go on for all of us."

And the voters certainly did make a choice.

On Wednesday, November 4 as voters from Crookston to Canton read the headlines in the morning paper, disbelief danced in their heads. The headlines couldn't be true, Jesse Ventura was elected governor.

The same guy who used to wear a pink boa as he spewed faux death threats on national television would be in charge of state spending priorities and the Minnesota National Guard. The same talk-show host who drove a Porsche and smoked big, fat cigars would be living in the governor's mansion. The same bald-headed, barrel-chested guy who used to be called "The Body" would now be referred to as "Governor Body." Heck, Jesse Ventura wasn't even his real name. His real name was James Janos, but the Minneapolis wrestler changed it early in his career to sound more Californian.

Ventura's time in office wasn't quite as successful as his campaign. Governing a state is complex, especially when the legislature is in session. He traded the Porsche for a chauffeured sedan. He tolerated bodyguards, but joked, "It wouldn't be too good if the Governor knocks somebody down."

And he traded his sweatpants for suits.

"Suits used to take me forty minutes," Ventura said. "Now I can do it in twelve."

DEATH OF THE
AMERICAN CONSCIENCE

- 2002 -

OCTOBER 25 WAS LESS THAN TWO WEEKS BEFORE Election Day, and even though the polls showed Democratic Senator Paul Wellstone with a comfortable lead, he knew enough about running a race to know it was going to get tight. He was working as if he were running behind. What he didn't need was bad weather or to face his fear of flying in small planes. He got both.

Wellstone needed to get from the Twin Cities to Eveleth, and it would be a quick trip on a Beechcraft King A100 turboprop airplane. The weather in Eveleth, however, was bad and the pilot, Richard Conry, canceled the flight. But, when the weather started to clear, Conry changed his mind and decided to proceed with the flight.

Wellstone asked Conry to check the weather again, just one more time.

Conry did and the report came back that the clouds were lifting and the freezing rain conditions were also improving. At that time, another charter pilot just returning from Duluth, not far from the Eveleth-Virginia airport, said he hadn't run into any trouble.

Conry made the quick flight to the region and was cleared by air traffic controllers to land at the Eveleth-Virginia Municipal Airport. The plan was for the plane to pick up an electronic signal that would guide the plane through the clouds to the end of the runway, where it could then make a safe landing. But Conry and copilot Michael Guess missed the first chance to intercept the signal correctly and had to turn the plane to pick up the approach properly.

After a second attempt failed, Conry and Guess tried one last time. On this final attempt the plane picked up too much speed, which delayed the landing gear. In response to that, the pilots slowed the plane down. According to reports, the pilots were required to keep the plane at a set speed until the runway was in sight.

But the clouds were dense and hung low to the ground. Both pilot and copilot struggled to see the ground or the runway. The rain drizzled and the day turned gray. As they looked for the runway, both men at the controls failed to notice that the speed gauge had fallen below eighty-five knots. The plane stalled, then plummeted to the ground, just a couple miles short of the airport. Wellstone, his wife and daughter, three of the senator's staff, and both pilots were killed.

While many at first suspected icy, foggy conditions as the reasons for the crash, the National Transportation Safety Board (NTSB) ruled that pilot error was the cause.

The safety board also found that Conry had "a record of below-average performance," with many complaints about his work as a pilot. Ironically, Conry had passed a proficiency check just two days before the fatal crash.

Other factors about Conry that may have contributed to the crash began over a day before his flight with Wellstone.

Conry had been a pilot on an emergency medical flight to North Dakota in the early morning hours of October 24. He drove back to his Minnetonka home at 9:30 that morning, twenty-four hours before he was scheduled to take off with Wellstone. Instead of sleeping at home that day, Conry went to work late in the afternoon at a second job as a dialysis nurse, which the charter plane company said it never knew of. What role pilot fatigue might have played in the crash is a matter of speculation.

News of the crash also brought on immediate conspiracy theories, since Wellstone's Senate seat could disrupt the delicate balance of power in the Senate. Many Web postings, a few articles, even a book built a case, not convincing to some, that Wellstone's plane crash was an assassination, not accidental. Some accused President George W. Bush's administration of being behind Wellstone's death because Wellstone was one of the administration's most strident critics, especially on the Iraq War. One researcher pointed out that Democrats are twice as likely to die in plane crashes as Republicans. Some conspiracies centered around some sort of sabotage that disabled the pilot's navigation system. One person noted odd cell phone behavior in the area at the time of the crash and pointed out that there was no distress call from the plane.

While theories persist on the Web, they've been largely debunked by the NTSB's findings and other officials.

On October 29 as many as twenty thousand people jammed into the Williams Arena on the University of Minnesota campus to pay tribute to Wellstone. What was supposed to be a funeral service and a memorial turned into a political rally that obliterated any chance of the Democrats holding on to Wellstone's Senate seat, even though former vice president Walter Mondale agreed to run in the fallen senator's place.

JumboTron screens were used at the service, and the funeral was carried live on national television. The atmosphere quickly became like that of a sporting event. When prominent Republicans who came to pay their respects were shown on the screens, the crowd booed.

The crowds' mournful exuberance could have possibly been overlooked if not for Rick Kahn, one of Wellstone's protégés and a longtime friend. In many respects, Kahn was the opposite of Wellstone, shy and normally quiet. But in the wake of the tragedy, Kahn's memorial speech was a fiery sermon. For twenty-five minutes Kahn waxed political in a rapturous tone which the crowd devoured. Meanwhile, politicians and others tuning in were turned off. Minnesota governor Jesse Ventura left the ceremony furious. Other politicians felt duped into attending a political rally.

Kahn told supporters, "We are begging you to help us win this election for Paul Wellstone."

He then called out certain Republicans in the audience, suggesting that the Grand Old Party should concede the election out of respect.

"We can redeem the sacrifice of his life if you help us win this election for Paul Wellstone," Kahn said.

The speech did not have the desired affect and the resulting polls showed it. Norm Coleman, Wellstone's opponent and former mayor of St. Paul began to seize on the momentum. Coleman, who had been languishing in the polls, began to make up ground. Mondale's eleven-day Senate run was doomed.

Wellstone's campaign manager apologized for the event, even though Kahn didn't.

"I didn't say anything bad at all, and I wouldn't. It's not in my heart," Kahn said.

The memorial service, not Wellstone's legacy or the war in Iraq, became the defining issue of the campaign, and the Republicans oddly had become the martyrs.

A *Time* magazine poll showed that 49 percent of the voters in Minnesota were less likely to vote Democrat after the rally. And two-thirds said the funeral service made them feel less positive.

It was a sad epilogue to a brilliant Senate career that started with a college professor traversing the state in a green school bus.

Wellstone remained true to his vision and true to words that he spoke at Swathmore College in 1998: "I do not believe the future will belong to those who are content with the present, I do not believe the future will belong to the cynics, or to those who stand on the side-line. The future will belong to those who have passion, and those who are willing to make the personal commitment to make our country better. The future will belong to those who believe in the beauty of their dreams."

THE SPIRIT THAT WOULDN'T DIE:
GRAIN BELT BEER

- 2002 -

It was summer and it was pouring rain in St. Paul as the bottles of Grain Belt beer made a *clickety-clank* along the ancient production line. Most employees had learned to ignore the leaky roof, but it was harder to ignore when the machinery stopped. Employees of the Minnesota Brewing Company looked up at the clock—it was only noon.

Three hours later the packaging department finished loading the last of the golden-colored brew. Then they heard the call for a company meeting. Because of the rain and the leaky roof, the meeting was held in the rathskellar, a formal lounge for entertaining in the basement. For several months, newspapers had speculated about the brewing company's financial problems. But now the news was confirmed, and nearly a hundred employees vacated the plant to tell their families they were unemployed. The parking lot at the brewery was cleared within thirty minutes.

Glass bottles were left sitting on the production line, waiting to be filled by beer that wouldn't be brewed. Cans had been left in the pasteurizer.

Later that summer Schell Brewing Company of New Ulm, Minnesota, bought the Grain Belt brand name. By October, the brewery held a moving party to relocate the brew with the original recipe to its New Ulm facility.

The brewery expanded in order to produce the newly acquired Grain Belt label. Ironically, the beer that seemed to curse other companies in the past, helped Schell prosper in a time of increased pressure from names like Miller, Coors, and Budweiser.

Grain Belt had become the beer—possibly the only beer—to die and come back to life four different times and in three different Minnesota locations.

Grain Belt's arrival in Minnesota predates the state of Minnesota. In the 1850s, a French immigrant named John Orth started a brewing company near the Mississippi River. Conveniently, Orth used the cold caves on Nicollet Island to store the brew. By 1890 Orth had merged his brewing business with three other Minneapolis brewers, incorporating their company as Minneapolis Brewing Company. In 1893 the company began calling its beer "Grain Belt."

Slowly the reputation of the brew grew beyond Minneapolis, and the state developed a taste for Grain Belt.

Grain Belt missed its seventy-fifth birthday because of the Volstead Act, also known as Prohibition. Luckily, the Minneapolis Brewing Company saw Prohibition coming when Minnesota passed the County Options Law in 1915, which allowed counties to decide whether to allow alcohol on a county-by-county basis. Minneapolis Brewing Company started a subsidiary company, the Golden Grain Juice Company, which manufactured nonalcoholic drinks. That company would keep Grain Belt alive despite the beer's Prohibition banishment. Golden Grain Juice Company rolled out barrels of near-beer called Minnehaha Pale. It contained less than one-half of 1 percent alcohol. During Prohibition the brewery was retrofitted with stills, a de-alcoholizer, and kept under the paternal control of the revenuers.

In 1933 Grain Belt reappeared when the Volstead Act disappeared in the midst of a great economic depression. Minnesota's friendly brew enjoyed a long period of prosperity, solidifying itself as the number-one selling beer in Minnesota for years.

In the mid-1970s, the beer landscape began to change dramatically. Regional breweries like Grain Belt began to feel pressure from national breweries like Anheuser-Busch and Miller, which had poured millions of dollars into advertising and country-wide distributing and slashed costs in order to open up new markets. Grain Belt struggled to keep up.

Minnesota financier Irwin Jacobs bought the struggling company, held onto it for eight months, than liquidated the company leaving thousands of Grain Belt fans uncertain if they were drinking the last of the "friendly beer." In November 1975, Jacobs sold Grain Belt to G. Heileman Brewery. Heileman terminated almost four hundred employees. And, to add insult to injury, Heileman changed Grain Belt's recipe and featured the Wisconsin-based Heileman corporate logo on the cans, further alienating Minnesota beer drinkers. For many, Grain Belt had died a second time.

Grain Belt's fate for the next fifteen years would be tied with Heileman's. Heileman Brewery eventually fell into the hands of the Bond Corporation, and in 1990 that company reported losses of $1.86 billion. By September 1990 the brewery in St. Paul, along with the Grain Belt brand, had been put up for sale. Heileman had trouble finding a buyer for Grain Belt and shut the doors in July 1991.

Grain Belt appeared to be dead again.

Then, Bruce Hendry successfully bought the plant for $3.2 million. The brewery reopened in October 1992 as the Minnesota Brewing Company. Minnesota Brewing Company enjoyed success with Grain Belt and a new generation got its first taste of a Minnesota tradition. In 1993 Minnesota Brewing Company rolled out the hundred-year label, celebrating the centennial of the "Grain Belt" name, even though the beer itself had been brewed for much longer.

But, by the twenty-first century, Grain Belt had again fallen on hard times. The St. Paul plant faced huge debts, almost fifteen million dollars. Again, Grain Belt appeared to be dead.

For Minnesota beer lovers, August Schell Brewery would save the beer that they and their grandparents had loved.

MINNESOTA FACTS AND TRIVIA

Minnesota has 11,842 lakes to be exact; it just doesn't sound as catchy on a license plate.

Minnesota's state bird is the common loon. This red-eyed bird is known for its eerie falsetto calls. It swims underwater to catch fish and can dive more than ninety feet.

Minnesota—the land of ten thousand lakes—has four counties that have no natural lakes.

The Minnesota state flag has three dates woven into it. The largest, 1858, is the year Minnesota became a state. The one on the left-hand side is 1819, the year Fort Snelling was settled, essentially marking the first permanent European settlement in Minnesota. The final date, 1893, is the year the state flag was adopted.

Minnesota's capital, St. Paul, was originally named "Pig's Eye" after a whiskey trader named Pierre "Pig's Eye" Parrant, who first settled the town. The original Indians who lived in St. Paul called it "IM-IN-I-JA SKA," which means "white rock." It's not the only town in Minnesota that changed its mind before settling on a name. Just down the Mississippi River from St. Paul sits Winona, which was originally called "Montezuma." The city renamed itself Winona after the American Indian princess.

The name Minnesota comes from the Dakota Sioux word for "sky-tinted water," which refers to the many blue lakes in the state.

Walter Liggett, a firebrand journalist intent on promoting unpopular ideas of his day like socialism and an ardent critic of politicians who he thought were corrupt—including the governor—was shot dead in 1935.

The Minnesota Fair (known to Minnesota residents as the "Great Minnesota Get-together") has been a tradition since 1854, when the state was still a territory. However, it has been canceled five times: in 1861 because of the Civil War; in 1862 because of the Dakota Sioux War; in 1893 because of the World's Columbian Exposition not too far away in Chicago; in 1945 because of the shortage of fuel during World War II; and in 1946 because of a polio outbreak.

Congressman Andrew Volstead of Granite Falls, Minnesota, helped lead the charge against liquor in the first part of twentieth century. He authored the bill and an act that would later become the Eighteenth Amendment of the U.S. Constitution, outlawing liquor and ushering in Prohibition.

Pillsbury, a Minnesota company, unleashed Poppin' Fresh, the spokesdoughperson of the organization. Poppin' Fresh popped onto television in 1965 quipping, "Nothin' says lovin' like something from the oven." And people have been making him giggle ever since.

The official drink of Minnesota is milk. Minnesota ranks number six in milk production and produces about 6 percent of the nation's milk supply.

Minnesota is the birthplace of authors F. Scott Fitzgerald and Sinclair Lewis. Fitzgerald's father was a wicker furniture salesman, and Lewis's book *Main Street* was banned because Alexandria, Minnesota, thought the fictional town in the book was a thinly veiled reference to it.

The Honeycrisp apple is the state fruit of Minnesota. It was adopted in 2006 and is the brainchild (or fruitchild) of the University of Minnesota's apple program. The goal was to develop a tree that could survive winter and still produce high-quality fruit. The original seedling—a cross between a Macoun and Honeygold—was planted in 1962.

NBA Hall of Famer Kevin McHale, former baseball star Roger Maris, and former Charles Manson prosecutor Vincent Bugliosi were all born in Hibbing, Minnesota.

Minnesota actually has a state grain—wild rice, which was a staple for American Indians in Minnesota. Although its name connotes a similarity to rice, it's actually an aquatic grass. The plants bloom in early summer and by late summer the seeds mature into dark brown kernels. Most of the world's wild rice still comes from Minnesota. It's harvested from lakes by means of a canoe. Those wanting to harvest wild rice must get a license, just like hunting or fishing.

There are 201 mud lakes, 154 long lakes, and 123 rice lakes in Minnesota.

Before Richard Warren Sears, of Sears Roebuck fame, got started in retail, he made a living selling watches to train agents along the Minneapolis and St. Louis railway from Redwood Falls, Minnesota.

The stapler was invented in Spring Valley, Minnesota.

Before the Vikings moved to Minnesota, the NFL had two football teams in Duluth. One was the Kelley Duluths, and the other was the Eskimos.

Minnesota has over ninety thousand miles of shoreline, more than Hawaii, Florida, and California put together.

Most people know that Paul Bunyan, a Minnesota tall tale, had Babe the Blue Ox. What most people don't know is that Babe had a mate—Bessie the Yaller Cow.

When Tiny Tim had finally had enough tiptoeing through the tulips, he was buried in Lakewood Cemetery with his trusty ukulele, six tulips, and a stuffed rabbit.

The world's largest ball of twine is housed in Darwin, Minnesota. It was created by Francis Johnson and weighs nearly nine tons.

The first pop-up toaster was sold in Minnesota. In 1926 it cost $13.50.

BIBLIOGRAPHY

Hennepin's Wild Adventures—1680

Caruso, John Anthony. *The Mississippi Valley Frontier.* Indianapolis, Ind.: Bobbs-Merrill, 1966.

Hennepin, Louis. *A Description of Louisiana.* Ann Arbor, Mich.: University Microfilms, 1966.

———. *A New Discovery of a Vast Country in America, Volumes I and II.* New York: Kraus Reprints, 1972.

Myers, Ann K. D. "Was Hennepin County Named for a Charlatan?" James Ford Bell Library. http://bell.lib.umn.edu/hennepin/hennepin.html (accessed March 11, 2006).

Willis, John W. "Louis Hennepin." The Catholic Encyclopedia, 2006. www.newadvent.org/cathen/07215c.htm (accessed March 11, 2006).

Pike's Forgotten Discovery—1806

Hollon, Eugene W. *The Lost Pathfinder: Zebulon Montgomery Pike.* Norman: University of Oklahoma Press, 1949.

Moore, Bob. "Zebulon Pike: Hard Luck." U.S. National Park Service. www.nps.gov/jeff/zebulon_pike.html (accessed March 29, 2006).

————. "Zebulon Pike: Hard Luck Explorer or Successful Spy?" U.S. National Park Service. www.nps.gov/jeff/LewisClark2/Circa 1804/WestwardExpansion/EarlyExplorers/ZebulonPike.htm (accessed March 20, 2006).

Pike, Zebulon. *Sources of the Mississippi and the Western Louisiana Territory.* Ann Arbor, Mich.: University Microfilms, 1966.

Post, Tim. "Revisiting Zebulon Pike's Expedition to Minnesota." Minnesota Public Radio. http://news.minnesota.publicradio.org/features/2005/11/14_postt_zebpike (accessed March 29, 2006).

Pipestone—1831–1836

Institute of American Indian Studies. "Treaty with the Yankton Sioux, 1858." Institute of American Indian Studies, 2002. www.usd.edu/iais/siouxnation/treaty1858.html (accessed February 3, 2006).

McLeod, Christopher. "Pipestone." Earth Island Institute. www.sacredland.org/historical_sites_pages/pipestone.html (accessed February 4, 2006).

Murray, Robert A. *Pipestone: A History of Pipestone National Monument Minnesota.* Pipestone, Minn.: Pipestone Indian Shrine Association, 1965.

Museum of Nebraska Art. "George Catlin." Museum of Nebraska Art. http://monet.unk.edu/mona/artexplr/catlin/catlin.html (accessed April 3, 2006).

Navaille, Pat. "Pipestone National Monument." U.S. National Park Service, 2003. www.nps.gov/pipe/history.htm (accessed April 3, 2006).

Southwick, Sally J. *Building on a Borrowed Past: Place and Identity in Pipestone, Minnesota.* Athens: Ohio University Press, 2005.

Instant City: Just Add Residents—1852

Chapan, Earl. "A Tragic City That Never Happened." *St. Paul Pioneer Press,* March 23, 1947.

Christenson, Jerome. *City Mouse–Country Mouse: An Illustrated History of Winona County, Minnesota.* Virginia Beach, Va.: Donning, 2005.

Curtiss-Wedge, Franklyn. *A History of Winona County.* Chicago: H.C. Cooper Jr., 1913.

Winona Republican-Herald. "Colony at Minnesota City Once Largest in the State." Sec. A, November 30, 1930.

———. "Minnesota City in Early Days Held Place Ahead of Winona." n.d.

Robbing St. Peter to Pay St. Paul—1857

Aby, Anne J., ed. *The North Star State: A Minnesota History Reader.* St. Paul: Minnesota Historical Society Press, 2002.

Blegen, Theodore C. *Building Minnesota.* Boston: D.C. Heath, 1938.

———. *Minnesota: A History of the State.* Minneapolis: University of Minnesota Press, 1963.

Christianson, Theodore. *Minnesota: The Land of Sky-Tinted Waters, a History of the State and Its People.* Chicago: American Historical Society, 1935.

Lass, William E. *Minnesota: A Bicentennial History.* New York: Norton, 1977.

Roethke, Leigh S. *Minnesota's Capitol: A Centennial Story.* Afton, Minn.: Afton Historical Society Press, 2005.

Dakota Sioux War—1862

Bessler, John D. *Legacy of Violence: Lynch Mobs and Executions in Minnesota.* Minneapolis: University of Minnesota Press, 2003.

Blegen, Theodore C. *Building Minnesota.* Boston: Heath, 1938.

———. *Minnesota: A History of the State.* Minneapolis: University of Minnesota Press, 1963.

Christianson, Theodore. *Minnesota: The Land of Sky-Tinted Waters, a History of the State and Its People.* Chicago: American Historical Society, 1935.

Lass, William E. *Minnesota: A Bicentennial History.* New York: Norton, 1977.

The Largest Execution in U.S. History—1862

Bessler, John D. *Legacy of Violence: Lynch Mobs and Executions in Minnesota.* Minneapolis: University of Minnesota Press, 2003.

Blegen, Theodore C. *Building Minnesota.* Boston: Heath, 1938.

———. *Minnesota: A History of the State.* Minneapolis: University of Minnesota Press, 1963.

Christianson, Theodore. *Minnesota: The Land of Sky-Tinted Waters, a History of the State and Its People.* Chicago: American Historical Society, 1935.

Lass, William E. *Minnesota: A Bicentennial History.* New York: Norton, 1977.

St. Paul Pioneer Press. "Execution." Sec. A, December 27, 1862.

The Founding of the Mayo Clinic—1865

Braasch, William F. *Early Days in the Mayo Clinic.* Springfield, Ill.: Charles C. Thomas, 1969.

Clapesattle, Helen. *The Doctors Mayo.* Minneapolis: University of Minnesota Press, 1941.

Wilder, Luch. *The Mayo Clinic.* Minneapolis, Minn.: McGill, 1936.

Watkins at Your Front Door—1868

Christenson, Jerome. "Pieces of the Past: Celebrating Winona's First 150 Years." *Winona Daily News,* 2001.

Goplen, John. *Images of America: Watkins.* Charleston, S.C.: Arcadia, 2004.

Schmidt, Stan. "Watkins: Today and Tomorrow . . . the Business of the '90s." *Winona Daily News,* 1993.

Frozen Stiff: The Minnesota Blizzard—1873

Christianson, Theodore. *Minnesota: The Land of Sky-Tinted Waters, a History of the State and Its People.* Chicago: American Historical Society, 1935.

Minneapolis Tribune. "Frozen to Death." Sec. A, January 11, 1873.

———. "The Great Storm." Sec. A, January 12, 1873.

———. "The Great Storm." Sec. A, January 14, 1873.

———. "The Mortuary Record." Sec. A, January 17, 1873.

———. "Snow-Flakes." Sec. A, January 9, 1873.

———. "The Storm." Sec. A, January 9, 1873.

———. "The Storm Described." Sec. A, January 15, 1873.

Potter, Merle. *101 Best Stories of Minnesota.* Minneapolis, Minn.: Schmitt Publications, 1956.

Rolvaag, O. E. *Giants in the Earth.* New York: Harper and Brothers, 1929.

Winona Republican. "After the Storm." Sec. A, January 10, 1873.

———. "Deaths from the Storm." Sec. A, January 14, 1873.

———. "The Railroads." Sec. A, January 11, 1873.

———. "The Storm." Sec. A, January 8, 1873.

———. "The Storm." Sec. A, January 9, 1873.

The Wild, Wild West of Minnesota—1876

Christianson, Theodore. *Minnesota: The Land of Sky-Tinted Waters, a History of the State and Its People.* Chicago: American Historical Society, 1935.

Craddock, Van. "Cole Younger Had an Arresting Personality." *Longview News-Journal,* Sec. A, October 15, 2005.

Folwell, William Watts. *A History of Minnesota.* St. Paul: Minnesota Historical Society, 1969.

Holmes, Frank R. *Minnesota in Three Centuries: 1655–1908.* Mankato: Publishing Society of Minnesota, 1908.

Minneapolis Tribune. "Northfield's Sensation." Sec. A, September 8, 1876.

Parker, Dick. "Retro: Old Outlaws in a New Century." *Star Tribune,* Sec. B, July 9, 2006.

Potter, Merle. *101 Best Stories of Minnesota.* Minneapolis, Minn.: Schmitt Publications, 1956.

Smith, Robert Barr. *The Last Hurrah of the James-Younger Gang.* Norman: University of Oklahoma Press, 2005.

St. Paul Dispatch. "The Dead Bandits." Sec. A, September 9, 1876.

Winona Republican. "The Northfield Tragedy." Sec. A, September 8, 1876.

The End of an Era: Minnesota Is Settled—1878

Caron, John. "Red River Carts." Institute for Regional Studies, North Dakota State University. www.fargo-history.com/transportation/red-river-carts.htm (accessed May 1, 2006).

City of Coon Rapids. "Red River Oxcart Trail." City of Coon Rapids, Minnesota. www.ci.coon-rapids.mn.us/council/Commissions/Historicalmural/oxcart.htm (accessed May 1, 2006).

"A Few Thoughts about the Red River Carts." *CCHS Newsletter,* March–April 2000. www.info.co.clay.mn.us/History/red_river_carts.htm (accessed May 1, 2006).

Gilman, Rhoda R., Carolyn Gilman, and Deborah Miller. *The Red River Trails.* St. Paul: Minnesota Historical Society Press, 1979.

Burning Down the House: The State Capitol Burns—1881

Christianson, Theodore. *Minnesota: The Land of Sky-Tinted Waters, a History of the State and Its People.* Chicago: American Historical Society, 1935.

Holmes, Frank R. *Minnesota in Three Centuries: 1655–1908.* Mankato: Publishing Society of Minnesota, 1908.

Minneapolis Tribune. "In Ashes." Sec. A, March 2, 1881.

———. "Legislative Acts." Sec. A, March 4, 1881.

Roethke, Leigh S. *Minnesota's Capitol: A Centennial Story.* Afton, Minn.: Afton Historical Society Press, 2005.

Winona Republican. "The Capitol Disaster." Sec. A, March 3, 1881.

———. "Destruction of State Capitol by Fire." Sec. A, March 2, 1881.

The Town So Nice It Burned Twice—1893 and 1900
Minneapolis Tribune. "Desolation: The Town of Virginia Burned by Flames." Sec. A, June 19, 1893.

———. "Little Left of Virginia." Sec. A, June 9, 1900.

———. "Rain: It Promises to Relieve the Situation in the Mesabi Range." Sec. A, June 22, 1893.

———. "Terrible: The Latest Intelligence from the Mesabi Range." Sec. A, June 20, 1893.

———. "Virginia, Minn., Is Again Destroyed by Fire." Sec. A, June 8, 1900.

———. "Virginia: The Town Wiped out of Existence by Fire." Sec. A, June 20, 1893.

Works Projects Administration in Minnesota. *The Minnesota Arrowhead Country.* Chicago: Albert Whitman, 1941.

Two Trains to Salvation: The Hinckley Fire—1894
Folwell, William Watts. *A History of Minnesota.* St. Paul: Minnesota Historical Society, 1969.

Holmes, Frank R. *Minnesota in Three Centuries: 1655–1908.* Mankato: Publishing Society of Minnesota, 1908.

Macalester College Geography Department. "The Story of the Hinckley Fire." Macalester College. www.macalester.edu/geography/mage/urban/hinckley/fire.htm (accessed October 5, 2006).

New York Times. "Burned in the Morass." Sec. A, September 4, 1894.

———. "The Dead Lie in Heaps at Hinckley." Sec. A, September 4, 1894.

———. "A Passenger Train Missing." Sec. A, September 3, 1894.

———. "Race for Life with Flames." Sec. A, September 3, 1894.

Potter, Merle. *101 Best Stories of Minnesota*. Minneapolis, Minn.: Schmitt Publications, 1956.

Swenson, Grace Stageberg. *From the Ashes: The Story of the Hinckley Fire of 1894*. Stillwater, Minn.: Croixside Press, 1979.

The Discovery of the Kensington Rune Stone—1898

Blegen, Theodore C. *The Kensington Rune Stone: New Light on an Old Riddle*. St. Paul: Minnesota Historical Society, 1968.

Landsverk, O. G. *Ancient Norse Messages on American Stones*. Glendale, Calif.: Norseman Press, 1969.

———. *The Kensington Runestone*. Glendale, Calif.: Church Press, 1961.

Meier, Peg. "Farmer's Not a Fraud, Family Says." *Star Tribune*, Sec. B, September 6, 2004.

———. "The Idea That Wandering Norseman Reached Minnesota in 1362 and Left Behind . . . " *Post-Gazette* (Pittsburgh, Pa.), Sec. B, December 4, 2000.

———. "Latest Runestone Tale: Second Find Is a Fake." *Star Tribune,* Sec. A, November 6, 2001.

———. "Runestone Takes Some More Lumps." *Star Tribune,* Sec. A, April 8, 2004.

———. "Smithsonian's Second Opinion: Runestone Is a Fake." *Star Tribune,* Sec. B, November 30, 2002.

Munier, Diane. "Minnesota's Runestone Is the Real Thing." *Ottawa Citizen,* Sec. A, May 12, 2002.

Potter, Merle. *101 Best Stories of Minnesota.* Minneapolis, Minn.: Schmitt Publications, 1956.

Wahlgren, Erik. *The Kensington Stone: A Mystery Solved.* Madison: University of Wisconsin Press, 1958.

Black Mass—1915

Christenson, Jerome. "Pieces of the Past: Celebrating Winona's First 150 years." *Winona Daily News,* 2001.

Crozier, William I. *Gathering a People: A History of the Diocese of Winona.* Winona, Minn.: Saint Mary's Press, 1989.

Winona Republican-Herald. "Bishop's Condition Shows Improvement." Sec. A, August 28, 1915.

———. "Bishop Heffron Shot Twice." Sec. A, August 27, 1915.

———. "Father Lesches Sent to Asylum at Saint Peter." Sec. A, December 3, 1915.

Walking on Water: Waterskiing Is Invented—1922

Brown, Curt. "He Walked the Plank in Minnesota." *Star Tribune,* Sec. C, July 6, 1993.

Gustafson, Paul. "Seventy-Five Years of Water-Skiing." *Star Tribune,* Sec. B, June 30, 1997.

Reusse, Patrick. "Legends of Summer." *Star Tribune,* Sec. C, July 25, 1999.

Ziemer, Gregor. *A Daredevil and Two Boards.* Madison, Wis.: Hunter Halvorson, 2005.

The Birth of Betty Crocker—1924

Chianello, Joanne. "Searching for Betty Crocker." *Ottawa Citizen,* Sec. I, May 7, 2005.

Marks, Susan. *Finding Betty Crocker: The Secret Life of America's First Lady of Food.* New York: Simon and Schuster, 2005.

Nelson, Rick. "The Betty Behind the Brand." *Star Tribune,* Sec. T, May 12, 2005.

Out of This World: The Milky Way in Minnesota?—1924

Brenner, Joel Glenn. "A Candy Bar King Bar None." *Washington Post,* Sec. E, July 6, 1999.

―――. *The Emperors of Chocolate: Inside the Secret World of Hershey and Mars.* New York: Random House, 2000.

Kettle, Martin. "The Candy Man." *Guardian,* Sec. T, July 6, 1999.

National Post. "Candy Bar Inventor Was Not a Sweet Man: Enormously Wealthy, Attended High School in Lethbridge, Alta." Sec. A, July 6, 1999.

Parker, Dick. "Minnesota Made—Candy Connection." *Star Tribune,* Sec. B, November 16, 2004.

Pottker, Jan. *Crisis in Candyland: Melting the Chocolate Shell of the Mars Family Empire.* Bethesda, Md.: National Press Books, 1995.

Reed, Christopher. "Forrest Mars." *Guardian,* Sec. A, August 10, 1999.

Richardson, Tim. *Sweets: A History of Candy.* New York: Bloomsbury, 2002.

The Milford Mine Tragedy—1924

Aulie, Berger. *The Milford Mine Disaster: A Cuyuna Range Tragedy.* Virginia, Minn.: W.A. Fisher, 1994.

Crosby Courier. "Progressing at Milford Mine." Sec. A, March 28, 1924.

Duluth Herald. "Frightful Disaster at Iron Mine Near Crosby." Sec. A, February 7, 1924.

Tice, D. J. *Minnesota's Twentieth Century: Stories of Extraordinary Everyday People.* Minneapolis: University of Minnesota, 2001.

Minnesota's Oldest Residents—1931

Dawson, Jim. "Ancient Browns Valley Man Skeleton Becomes Focus of Clash over Values." *Star Tribune,* Sec. A, January 15, 1989.

Jenks, Albert Ernest. *Pleistocene Man in Minnesota.* Minneapolis: University of Minnesota Press, 1936.

Meier, Peg. "Missing Prehistoric Bones Found." *Star Tribune,* Sec. E, January 15, 1989.

Otter Tail County Historical Society. "Otter Tail County Probably Has the Oldest Resident of North America." Otter Tail County Historical Society. www.co.otter-tail.mn.us/history/county history_mnwoman.php (accessed November 13, 2006).

Rensberger, Bryce. "Bones' Analysis Clouds Migration Theories." *Anchorage Daily News,* Sec. A, April 17, 1997.

Wright, Karen. "Reconstructing the Ancient Humans of the New World: The Remains of Spirit Cave Man Indicate the Early Inhabitants of the Americas Were Quite Different from Modern Aboriginals." *Vancouver Sun,* Sec. A, June 3, 1999.

SPAMtastic: The Invention of SPAM—1936

Blundo, Joe. "More than Everything You Always Wanted to Know About SPAM." *Columbus Dispatch,* Sec. E, April 8, 1999.

Gimacgima, Craig. "Spam-in the Globe." *Honolulu Star-Bulletin,* Sec. A, September 21, 2005.

Glasner, Joanna. "The History of SPAM, and Spam." Wired News, May 26, 2001. www.wired.com/news/business/1,44111-0.html (accessed December 10, 2006).

Lee, Thomas. "The Spam Mystique: Spam Has Been Hormel's Signature Brand for 67 Years, and Experts Say That Success Is Due to Hormel's Marketing Savvy." *Star Tribune,* Sec. D, November 19, 2004.

Wyman, Carolyn. *Spam: A Biography.* San Diego, Calif.: Harcourt Brace, 1999.

The Armistice Day Blizzard—1940

Anderson, Dennis. "River of Death." *Star Tribune,* Sec. C, November 8, 2000.

Christenson, Jerome. "Pieces of the Past: Celebrating Winona's First 150 years." *Winona Daily News,* 2001.

MacQuarrie, Gordon. "The Ducks Came, the Men Died." *Milwaukee Journal,* Sec. A, November 13, 1940.

Minnesota Historical Society. "Armistice Day Blizzard." Minnesota Historical Society. http://climate.umn.edu/doc/journal/top5/numbertwo.htm (accessed December 10, 2006).

National Weather Service. "Armistice Day Storm." National Weather Service, 2006. www.crh.noaa.gov/arx/events/armistice.php (accessed December 10, 2006).

Orloff, Katie. "Local Man Honored for His Heroics in 1940 Blizzard." *Press-Enterprise* (Riverside, Calif.), Sec. A, November 25, 1999.

Schara, Ron. "Blizzard Memories Still Haunt Hunters." *Star Tribune*, Sec. C, November 11, 1990.

Steil, Mark. "The Winds of Hell." Minnesota Public Radio, November 10, 2000. http://news.minnesota.publicradio.org/features/ 200011/10_steilm_blizzard-m (accessed December 10, 2006).

Winona Republican-Herald. "Hunters Trapped by Storm, Seven Die." Sec. A, November 12, 1940.

The Night That Music Lived—1959
Bream, Jon. "Bobby Vee's Tribute CD Recalls Holly 40 Years After His Death." *Star Tribune*, Sec. E, February 5, 1999.

Helm, Merry. "Day the Music Died. North Dakota Public Radio." February 3, 2006. www.prairiepublic.org/programs/datebook/ bydate/06/0206/020306.jsp (accessed December 27, 2006).

Huxley, Martin, and Quinton Skinner. *Behind the Music: The Day Music Died.* New York: Pocket Books, 2000.

Katz, Larry. "A Survivor of the Day Music Died." *Boston Herald*, Sec. A, November 9, 2000.

Lehmer, Larry. *The Day the Music Died: The Last Tour of Buddy Holly, the "Big Bopper" and Ritchie Valens.* New York: Schirmer Trade Books, 1997.

MacDonald, John. "Vee Recalls the Day After Music Died." *Milwaukee Journal Sentinel*, Sec. A, February 4, 1999.

Vee, Bobby. "Bobby Vee Biography." www.bobbyvee.com/bio.html (accessed December 10, 2006).

————. "Q&A." www.bobbyvee.com/interv.html (accessed December 10, 2006).

It's a Hit: The Homer Hanky—1987

Boswell, Thomas. "The Nice Guys Finished First." *Washington Post,* Sec. B, October 26, 1987.

Ehrlick, Darrell. Unpublished interview with Terrie Robbins. E-mail from Robbins to Ehrlick, April 9, 2007.

Klobuchar, Jim. "The Game Ball Bullies, Acts I and II." *Star Tribune,* Sec. B, July 3, 1991.

Oldenburg, Don. "Beyond the Wave." *Washington Post,* Sec. C, November 14, 1991.

Thornley, Stew. *Baseball in Minnesota.* St. Paul: Minnesota Historical Society Press, 2006.

Tordoff, Jeff. "It's Official: Homer Hanky a Historical Object." *Star Tribune,* Sec. A, October 22, 1987.

Tracy, Ben. "Good Question: What Is the Homer Hanky's History?" WCCO.com. http://wcco.com/goodquestion/local_story_277092044.html (accessed April 1, 2007).

The Ninth Wonder of the World—1992

Baenen, Laura. "Gawkers Cram the Halls as Huge Mall Debuts." *Houston Chronicle,* Sec. A, August 12, 1992.

Bentley, Rosalind. "Taking More than They Give." *Star Tribune,* Sec. A, August 5, 2002.

Blake, Laurie. "Fears of Gridlock Near Mall Aren't Borne Out." *Star Tribune,* Sec. A, August 12, 1992.

Bream, Jon. "But Where's My Car?" *Star Tribune,* Sec. A, August 11, 1992.

Diaz, Kevin, and Sally Apgar. "Heeding the Mall's Call." *Star Tribune,* Sec. A, August 12, 1992.

Farhi, Paul. "A Megamall Sprouts in the Minnesota Prairie." *Austin American Statesman,* Sec. D, August 11, 1992.

Grow, Doug. "From Wall to Wall, the Mall Will Be Everything but Small." *Star Tribune,* Sec. B, June 15, 1989.

Kennedy, Tony, and Sally Apgar. "The Doors Have Opened; Let the Shopping Commence." *Star Tribune,* Sec. A, August 11, 1992.

Levy, Melissa. "More Disneyland than Dales." *Star Tribune,* Sec. A, August 4, 2002.

Lowe, Jennifer. "Giant Mall's Debut Draws 150,000." *Orange County Register,* Sec. C, August 12, 1992.

Mall of America. "History of the Mall of America." Mall of America. www.mallofamerica.com/about_moa_history.aspx (accessed December 28, 2006).

———. "Mall of America Facts." Mall of America. www.mallof america.com/about_moa_mall_facts.aspx (accessed December 28, 2006).

Neely, Anthony. "Mega-Hopes Prevail at Mall Groundbreaking." *Star Tribune,* Sec. A, June 15, 1989.

Ramirez, Marc. "Breaking the Bank and Boggling the Mind." *Seattle Times,* Sec. K, July 2, 2006.

Tutelian, Louise. "In Minnesota, a Mall as Big as All Indoors." *New York Times,* Sec. F, November 17, 2006.

Wieffering, Eric. "Mall Stands Alone." *Star Tribune,* Sec. A, August 4, 2002.

Woutat, Donald. "It's so Big That . . . " *Newsday,* Sec. A, August 11, 1992.

The Ragin' Red: The Flood of East Grand Forks—1997

Broderson, Michelle. "The Flood of the Century." Michigan Technical University. www.geo.mtu.edu/department/classes/ge404/mlbroder (accessed January 9, 2007).

Loyola University. "The Anatomy of the Mississippi River." Loyola University. www.americaswetlandresources.com/background_facts/detailedstory/MississippiRiverAnatomy.html (accessed January 9, 2007).

Pielke, Roger A., Jr. "Who Decides: Forecasts and Responsibilities in the 1997 Red River Flood." *Applied Behavioral Science Review* 17 (1991): 83–101.

Quam, Jennifer. *A Small Town's War: East Grand Forks 1997 Flood Fight.* East Grand Forks: City of East Grand Forks Minnesota, 1999.

Shelby, Ashley. *Red River Rising: The Anatomy of a Flood and the Survival of an American City.* St. Paul, Minn.: Borealis Books, 2003.

The Body Becomes the Governor—1998

Associated Press. "'The Body' Becomes the Guv." *Florida Times Union*, Sec. A, January 5, 1999.

Belluck, Pam. "A Tough Match: 'Jess the Gov.' vs. Daily Grind." *New York Times*, Sec. A, January 4, 1999.

Fischer, Marc. "The Body Politic." *Grand Rapids Press*, Sec. A, November 4, 1999.

Groeneveld, Benno. "Agenda Slim as 'The Body' Becomes the Governor: Former Wrestler Ventura Promises He'll Tell the Truth." *National Post*, Sec. A, November 6, 1996.

Hauser, Tom. *Inside the Ropes with Jesse Ventura*. Minneapolis: University of Minnesota Press, 2002.

Klobuchar, Jim. "Wrestling with a New Political Era." *Christian Science Monitor*, Sec. A, November 6, 1998.

Kuntzman, Gersh. "'The Body' Will Keep His Head." *New York Post*, Sec. A, January 4, 1999.

Lentz, Jacob. *Electing Jesse Ventura: A Third Party Success Story*. Boulder, Colo.: Lynne Reinner, 2002.

Overholser, Geneva. "Ventura: Just Like One of Us." *Sunday Gazette-Mail* (Charleston, W.Va.), Sec. B, February 28, 1999.

Rubenstein, Steve. "Jesse Hard to Pin." *San Francisco Chronicle*, Sec. A, November 6, 1998.

Schultze, Steve. "'The Body' Becomes 'The Governor.'" *Milwaukee Journal Sentinel*, Sec. A, November 5, 1998.

St. Anthony, Neal. "Ventura Catches Big Business Off Guard." *Star Tribune,* Sec. D, November 6, 1998.

Death of the American Conscience—2002

CNN.com. "Key Clues Lost in Wellstone Crash." CNN.com, October 29, 2002. http://archives.cnn.com/2002/US/Midwest/10/28/wellstone.crash.

Diaz, Kevin. "Findings Don't Slow Conspiracy Theories on Wellstone Crash." *Star Tribune,* Sec. A, June 3, 2003.

Four Arrows and Jim Fetzer. *American Assassination: The Strange Death of Senator Paul Wellstone.* Brooklyn, N.Y.: Vox Pop, 2004.

Howe, Patrick. "Observers Examine How Wellstone Service Wounded Democrats." *St. Louis Post-Dispatch,* Sec. A, November 28, 2002.

Kennedy, Tony. "A Call for Change Arises from Wellstone Crash." *Star Tribune,* Sec. B, November 20, 2003.

————. "Logbook Offers Intriguing Details." *Star Tribune,* Sec. B, December 22, 2002.

————. "The Wellstone Crash: Loved Ones Find Little Solace in Report." *Star Tribune,* Sec. A, November 19, 2003.

————. "The Wellstone Plane Crash: Experts Suspect Fatal Stall." *Star Tribune,* Sec. A, December 29, 2002.

Lofy, Bill. *Paul Wellstone: The Life of a Passionate Progressive.* Ann Arbor: University of Michigan Press, 2005.

Shepard, Scott. "Mondale Likely to Run for Senate." *Atlanta Journal Constitution,* Sec. A, October 28, 2002.

Waid, Matthew L. "Pilots Blamed for Wellstone Crash." *New York Times,* Sec. A, November 13, 2003.

Zeleny, Jeff, and John McCormick. "Minnesota Senator Killed in Plane Crash with Wife, 5 Others." *Knight Ridder Tribune News Service,* October 25, 2002.

The Spirit That Wouldn't Die: Grain Belt Beer—2002
Feyder, Susan. "Schell's Game." *Star Tribune,* Sec. D, January 15, 2003.

Gordon, Jack. "Irv the Boat Maker." *Twin Cities Business,* October 2006.

Kennedy, Tony. "Minnesota Brewing Closes Plant." *Star Tribune,* Sec. A, June 25, 2002.

———. "Schell Wins Auction for Grain Belt Brand Name." *Star Tribune,* Sec. D, July 23, 2002.

Koeller, Paul D., and David H. DeLano. *Brewed with Style: The Story of the House of Heileman.* La Crosse: University of Wisconsin-LaCrosse, 2004.

Lonto, Jeff R. "Grain Belt History." August Schell Brewing, 2004. www.grainbelt.com/history.php (accessed September 17, 2006).

———. *Legend of the Brewery: A Brief History of the Minneapolis Brewing Heritage.* Minneapolis, Minn.: Z-7 Publishing, 1998.

INDEX

Nichols, Hermann, 63
Nicollet County
    Courthouse, 25
Nicollet Island, 138
Northfield, 47, 48
Northwest Company, 9

O
Ohman, Olof, 70, 71
Olin, Rollin, 32
Orth, John, 138
Oshawa, 45
Oshman's, 119
Owens, Pat, 126
Owsley, Doug, 96, 97
oxcarts, 53–55, 56–57

P
Parr, W. E., 74
patent medicines, 39–41
Pelican Rapids, 94
Pike, Zebulon, 6–10
Pioneer Press, 127
pioneers, 55–56
Pipestone, 11–15
Pitts, Charlie, 48–50, 51
Plainview, 39
Pope, John, 32, 33
Prairie La Crosse, 18
Prescott, Philander, 14

Prison Mirror, 52
Prohibition, 138–39
Puckett, Kirby, 112, 114–15

R
railroads, 45, 66–69
Ramsey, Alexander, 28, 31, 33
Rattling Runner, 31
Ravoux, Father Augustin, 30
Reardon, Jeff, 113
Reardon Hotel, 52
Red Middle Voice, 26
Red River, 122–27
Red River Valley, 53, 55, 123
reservations, 26–27
Retaliate in 98, 130
Richardson, J. P., 107, 108
Riggs, Stephen, 31–32
Robbins, Terrie, 111–12
Rochester, 36, 37–38
Rolette, Joseph "Jolly Joe,"
    22–25, 55–56
Rollingstone Valley, 18
Rum River, 3
Runs Against Something
    When Crawling, 26

S
Saint Mary's College, 73
Saint Peter Company, 23

ABOUT THE AUTHOR

Darrell Ehrlick is the editor of the *Winona Daily News* and has spent the last decade writing in places like Fargo, North Dakota; Logan, Utah; and Winona, Minnesota. He was born in Billings, Montana, and attended graduate school at Emory University in Atlanta, Georgia, and Bennington College, in Bennington, Vermont. He is an avid reader and baseball fan. He and his wife, Angie, keep an eye on Minnesota from their home right across the Mississippi River in Trempealeau, Wisconsin. Their lives are run by four very demanding cats, all of which supervised the writing of this book (in different shifts) from the author's lap.